Plankton Communities and Summertime Declines in Algal Abundance Associated with Low Dissolved Oxygen in the Tualatin River, Oregon

By Kurt D. Carpenter and Stewart A. Rounds

Prepared in cooperation with Clean Water Services

Scientific Investigations Report 2013–5037

U.S. Department of the Interior
U.S. Geological Survey

U.S. Department of the Interior
KEN SALAZAR, Secretary

U.S. Geological Survey
Suzette M. Kimball, Acting Director

U.S. Geological Survey, Reston, Virginia: 2013

For more information on the USGS—the Federal source for science about the Earth, its natural and living resources, natural hazards, and the environment, visit http://www.usgs.gov or call 1–888–ASK–USGS.

For an overview of USGS information products, including maps, imagery, and publications, visit http://www.usgs.gov/pubprod

To order this and other USGS information products, visit http://store.usgs.gov

Suggested citation:
Carpenter, K.D., and Rounds, S.A., 2013, Plankton communities and summertime declines in algal abundance associated with low dissolved oxygen in the Tualatin River, Oregon: U.S. Geological Survey Scientific Investigations Report 2013–5037, 78 p.

Contents

Contents— Continued

Figures

Figures— Continued

Figures— Continued

Tables

Conversion Factors, Datums, and Abbreviations and Acronyms

Conversion Factors

Inch/Pound to SI

Multiply	By	To obtain
Length		
inch (in.)	2.54	centimeter (cm)
foot (ft)	0.3048	meter (m)
mile (mi)	1.609	kilometer (km)
Area		
acre	0.001562	square mile (mi^2)
square mile (mi^2)	2.59	square kilometer (km^2)
acre-foot (acre-ft)	43,560	cubic feet (ft^3)
Flow rate		
foot per second (ft/s)	0.3048	meter per second (m/s)
foot per mile (ft/mi)	0.1894	meter per kilometer (m/km)
cubic foot per second (ft^3/s)	0.02832	cubic meter per second (m^3/s)
mile per hour (mi/h)	1.609	kilometer per hour (km/h)

SI to Inch/Pound

Multiply	By	To obtain
Volume		
cubic meter (m^3)	35.3147	cubic foot (ft^3)
liter (L)	0.2642	gallon (gal)
Mass		
microgram (µg)	0.000001	gram (g)
gram (g)	0.035273	ounce, avoirdupois (oz)
kilogram (kg)	2.2046	pound (lb)

Temperature in degrees Celsius (°C) may be converted to degrees Fahrenheit (°F) as follows:

$$°F=(1.8×°C)+32.$$

Temperature in degrees Fahrenheit (°F) may be converted to degrees Celsius (°C) as follows:

$$°C=(°F-32)/1.8.$$

Specific conductance is given in microsiemens per centimeter at 25 degrees Celsius (µS/cm at 25°C).

Concentrations of chemical constituents in water are given either in milligrams per liter (mg/L) or micrograms per liter (µg/L).

Algal biovolumes for phytoplankton are given in cubic micrometers, or microns (µm), per milliliter of sample volume ($µm^3$/mL).

Algal cell densities for phytoplankton are given in cells per milliliter of sample (cells/mL), and zooplankton densities are given in number of organisms per cubic meter (#/m^3).

Conversion Factors, Datums, and Abbreviations and Acronyms— Continued

Datums

Vertical coordinate information is referenced to the North American Vertical Datum of 1988 (NAVD 88).

Horizontal coordinate information is referenced to the North American Datum of 1983 (NAD 83).

Altitude, as used in this report, refers to distance above the vertical datum.

Abbreviations and Acronyms

Abbreviation or Acronym	Definition
ANN	Artificial Neural Network model
ANOSIM	analysis of similarity
BEST	Bio-Env Stepwise multivariate analysis
BOD	biochemical oxygen demand
Chl-a	phytoplankton chlorophyll-a
CWS	Clean Water Services
DO	dissolved oxygen
FNU	formazin nephelometric unit
JWC	Joint Water Commission
NTU	nephelometric turbidity unit
ODEQ	Oregon Department of Environmental Quality
ORWSC	USGS Oregon Water Science Center
PAR	photosynthetically active radiation
RM	river mile
SOD	sediment oxygen demand
SRP	soluble reactive phosphorus
SIMPER	similarity percentage analysis
TMDL	Total Maximum Daily Load
TVID	Tualatin Valley Irrigation District
USGS	U.S. Geological Survey
WWTF	wastewater treatment facility
99W	Highway 99 west

Plankton Communities and Summertime Declines in Algal Abundance Associated with Low Dissolved Oxygen in the Tualatin River, Oregon

By Kurt D. Carpenter and Stewart A. Rounds

Executive Summary

Phytoplankton populations in the Tualatin River in northwestern Oregon are an important component of the dissolved oxygen (DO) budget of the river and are critical for maintaining DO levels in summer. During the low-flow summer period, sufficient nutrients and a long residence time typically combine with ample sunshine and warm water to fuel blooms of cryptophyte algae, diatoms, green and blue-green algae in the low-gradient, slow-moving reservoir reach of the lower river. Algae in the Tualatin River generally drift with the water rather than attach to the river bottom as a result of moderate water depths, slightly elevated turbidity caused by suspended colloidal material, and dominance of silty substrates. Growth of algae occurs as if on a "conveyor belt" of streamflow, a dynamic system that is continually refreshed with inflowing water. Transit through the system can take as long as 2 weeks during the summer low-flow period. Photosynthetic production of DO during algal blooms is important in offsetting oxygen consumption at the sediment-water interface caused by the decomposition of organic matter from primarily terrestrial sources, and the absence of photosynthesis can lead to low DO concentrations that can harm aquatic life.

The periods with the lowest DO concentrations in recent years (since 2003) typically occur in August following a decline in algal abundance and activity, when DO concentrations often decrease to less than State standards for extended periods (nearly 80 days). Since 2003, algal populations have tended to be smaller and algal blooms have terminated earlier compared to conditions in the 1990s, leading to more frequent declines in DO to levels that do not meet State standards. This study was developed to document the current abundance and species composition of phytoplankton in the Tualatin River, identify the possible causes of the general decline in algae, and evaluate hypotheses to explain why algal blooms diminish in midsummer.

Plankton and water-quality sample data from 2006 to 2008 were combined with parts of a larger discrete-sample and continuous water-quality monitoring dataset and examined to identify patterns in water-quality and algal conditions since 1991, with a particular emphasis on 2003–08. Longitudinal plankton surveys were conducted in 2006–08 at six sites between river miles (RM) 24.5 and 3.4 at 2- to 3-week intervals, or 5–6 per season, and in-situ bioassay experiments were conducted in 2008 to examine the potential effects of wastewater treatment facility (WWTF) effluent and phosphorus additions on phytoplankton biomass and algal photosynthesis. Phytoplankton and zooplankton community composition, streamflow, and water-quality data were analyzed using multivariate statistical techniques to gain insights into plankton dynamics to determine what factors might be most tied to the abundance and characteristics of the phytoplankton assemblages, and identify possible causes of their declines.

The connection between low-DO events and algal declines was clearly evident, as bloom crashes were nearly always followed by periods of low DO. Algal blooms occurred each year during 2006–08, producing maximum chlorophyll-a (Chl-a) values in June or July generally in the range of 50–80 micrograms per liter (μg/L). Bloom crashes and absence of sufficient algal photosynthesis in mid- to late-summer contributed to minimum DO concentrations that were less than the State standard of 6.5 milligrams per liter (mg/L) based on the 30-day mean daily concentration, for 62–74 days each year. At times, the absolute minimum State standard (4 mg/L DO) also was not met. To learn more about why low-DO events occurred, specific algal declines during 2003–08 were scrutinized to determine their likely causal factors. From this information, a series of hypotheses were formulated and evaluated in terms of their ability to explain recent declines in algal populations in the river in late summer.

Meteorological, streamflow, turbidity, water temperature, and conductance conditions in the Tualatin River during the 2006–08 summer seasons were not atypical. Natural flow comprised the majority (70–80 percent) of flow each year during spring, but then reduced to 38–40 percent during midsummer when WWTF effluent—which contributed as much as 36 percent—and flow augmentation releases comprised a greater fraction of the flow. Summer 2008 was unusual, however, in the prolonged influence from the Wapato Lake agricultural area near Gaston in the upper part of the basin. The previous winter flooding and levee breach at Wapato Lake caused a much greater area of inundation. As a result, drainage from this area continued into July, much later than normal. A subsequent algal bloom in Wapato Lake then seeded the upper Tualatin River, and this drainage had a profound effect on the downstream plankton community. A large blue-green algae bloom developed—the largest in recent memory—prompting a public health advisory for recreational contact for about two weeks.

Algal growths and surface blooms are a common feature of the Tualatin River. Most of the dominant algae have growth forms and morphologies that are well suited for planktonic life, employing spines and gas vacuoles to resist settling, forming colonies, and producing mucilage (or toxins) to resist zooplankton grazing. In 2006–08, 143 algal taxa were identified in 117 main-stem samples; diatoms and green algae were more diverse than blue-green, golden, and cryptophyte algae, although these later groups sometimes dominated the overall volumetric abundance (biovolume). The most frequently occurring taxa, occurring in 97–99 percent of samples, were flagellated cryptophytes *Cryptomonas erosa* and *Rhodomonas minuta*. Other important algal taxa included centric diatoms *Stephanodiscus*, *Cyclotella*, and *Melosira* species and colonial green algae *Scenedesmus* and *Actinastrum*. These taxa comprised the majority of the algal biovolume during much of the growing season. A general seasonal trend in the phytoplankton assemblages was observed, with dominance by filamentous centric diatoms *Stephanodiscus* and *Melosira* in spring and early summer, and flagellated cryptophytes and green algae, particularly *Chlamydomonas* sp., in late-summer; or, in 2008, dominance by blue-green algae *Anabaena flos-aquae* and *Aphanizomenon flos-aquae* during the Wapato Lake bloom event.

There were 99 zooplankton taxa identified from the Tualatin River in 2006–08, composed primarily of cladocerans, copepods, and rotifers. A seasonal increase in zooplankton abundance was observed in early summer just as or shortly after the phytoplankton population began to increase, with populations growing to 15,000–120,000 organisms per cubic meter in the lower river. Zooplankton abundance showed a predictable and distinct longitudinal downstream increase, particularly downstream of Highway 99W (RM 11.6). Although grazing rates were not measured, the data suggest that, at times, zooplankton grazing may affect algal abundance and species composition

in the Tualatin River, with diatoms becoming relatively less abundant and flagellated cryptophytes and green algae relatively more abundant during periods when zooplankton densities were highest.

Multivariate statistical analyses identified soluble reactive phosphorus (SRP), natural flow, flow augmentation, and WWTF effluent as important factors influencing Tualatin River phytoplankton populations, with zooplankton density (particularly rotifers and copepods), specific conductance, chloride, and water temperature also having an important influence. Although SRP was highly correlated with the plankton communities, that correlation was likely the result of high or low algal activity (uptake) as SRP concentrations were often reduced to low levels during blooms. While previous studies have already established that phosphorus, among other factors such as flow, places a theoretical cap on the size of the phytoplankton population in the river, sometimes algal declines occur when SRP concentrations are apparently sufficient. To identify alternative causal factors, additional analyses were performed without SRP to focus on other water-quality parameters, zooplankton density, and flow factors. Considering data for all 3 years and including just those samples from the lower Tualatin River not affected by the 2008 Wapato Lake drainage event, three factors (percentage of reservoir flow augmentation, total natural flow, and rotifer density) best explained variations in the phytoplankton assemblages.

Analyses focusing on the possible causes of algal declines included the above multivariate analyses, scrutiny of 10 specific instances of declines in algal populations during 2003–08 including several bloom–crash sequences, and analyses of historic routine watershed monitoring data from Clean Water Services. Six factors were hypothesized to be important in causing bloom crashes or impeding blooms from rebounding in August: (1) light limitation from cloudy weather, (2) a reduction in the plankton inocula or "seed" entering the lower river from upstream sources, (3) increased summer streamflows, (4) changes in the dominant sources of flow as the percentage of flow augmentation and WWTF discharges have increased, (5) zooplankton grazing, and (6) low concentrations of bioavailable phosphorus (<0.015 milligram per liter). All of these hypotheses are supported in some fashion by the available data and statistical analyses. Zooplankton grazing, short-term declines in photosynthesis from cloudy weather, total flow as it affects residence time, and the dominant source of flow are primary factors responsible for the low-DO events caused by declines in algae in the lower Tualatin River during late summer.

Cloudy weather and increased turbidity are known to inhibit algal growth in the Tualatin River, and slight increases in turbidity in recent years may be a problem. Upstream sources of algae are critical in determining the characteristics and size of downstream populations, as illustrated by the Wapato Lake bloom in 2008, but more data are needed from upstream to fully define the importance of this connection. The sources of flow, through their differential contribution of plankton inocula (quality and amount), were, at times, important factors affecting phytoplankton populations. While SRP concentrations were often most highly correlated with phytoplankton species community, the bioavailability of phosphorus is still somewhat unknown and there are several sources to consider. Preliminary bioassay tests suggested that while treated wastewater effluent may stimulate algae at 30 percent concentrations, negative effects (or decreased stimulation) on Chl-*a* and DO production may occur at concentrations of 50 percent. Targeted data collection and future research will be needed to further understand the importance of these factors on Tualatin River phytoplankton.

While the data and analysis completed for this report provide insights into future research and monitoring that would be useful to continue, additional monitoring of turbidity, Chl-*a*, and plankton abundance and species composition in the upper part of the basin would enhance our understanding of plankton dynamics and factors affecting phytoplankton abundance in the lower river. Assessment of the key upstream sources of algal inocula via surveys of the major flow sources as well as tributaries and wetlands would provide useful information for the management of river water quality. Other studies that could prove useful for developing management strategies include targeted experiments to evaluate the bioavailability of phosphorus from a variety of sources. New research on phytoplankton–zooplankton interactions, and studies of planktivorous fish, might also provide insight about food web dynamics and potential "top-down" effects of fish predation on the plankton communities. In addition, further development of neural-network or other water-quality models would help to evaluate management strategies and provide forecasts of water-quality conditions. Finally, periodic future reassessments of the available data with the multivariate statistical tools used in this study would be helpful to assess whether and how plankton communities are changing, and to continue to shed light on the importance of factors shaping the plankton. Although certain types and sizes of algal blooms are undesirable, minimum phytoplankton populations are an important part of aquatic food webs and are needed to maintain healthy levels of DO in the river. By understanding the sources, characteristics, causal factors, and responses of the plankton communities, management strategies can be developed to improve DO conditions in the lower Tualatin River during the important summer low-flow period.

Introduction

The Tualatin River drains a 712-square-mile (mi^2) basin west of Portland in northwestern Oregon (fig. 1). Although the river descends from the Coast Range as a mountainous stream, it cuts and meanders as a relatively slow moving river across a fertile, fine-sediment filled valley for most of its 80-mile (mi) length. The Tualatin River has a history of algal blooms and associated water-quality problems dating back to the 1960s (table 1). At that time, blooms flourished in the sewage- and nutrient-enriched, warm, and slow-moving waters during summer. Algal blooms were still an aesthetic nuisance decades later, causing high pH and low concentrations of dissolved oxygen (DO) following bloom crashes, sometimes to levels that violated State of Oregon water-quality standards (Oregon Department of Environmental Quality, 2001). These problems are most severe in the pooled "reservoir reach" located from river mile (RM) 33.3 to RM 3.4, the end of which is marked by a low-head weir, the Oswego Dam (fig. 2). The average width in this reach increases to about 150 feet (ft) and the river becomes lake-like during the summer low-flow period, taking as long as 14 days for water to transit this slow-moving pool (Rounds and others, 1999).

The dominant types of algae in the Tualatin River during summer, based on a limited number of past studies, included planktonic diatoms (*Stephanodiscus*, *Cyclotella*, *Aulacoseira*, *Melosira*, and others), *Chlamydomonas* and other green algae, and occasional blooms of potentially toxic blue-green algae, including *Anabaena* and *Microcystis* (Carter and others, 1976; Rinella and others, 1981; Oregon Department of Environmental Quality and Unified Sewerage Agency, 1982; Doyle and Caldwell, 1996).

In 1972, the Oregon Department of Environmental Quality (ODEQ) listed the Tualatin River as "water-quality limited" because of impacts to aesthetics caused by algal blooms. In 1988, regulatory pollution limits, or Total Maximum Daily Loads (TMDLs), were established for ammonia and total phosphorus that were fueling excessive algal growth and causing problems associated with low DO and high pH levels (Oregon Department of Environmental Quality, 2001). Although nutrient controls in the early 1990s reduced the severity of algal blooms, phytoplankton Chl-*a* concentrations continued to exceed ODEQ's target upper limit of 15 µg/L to prevent nuisance algal conditions.

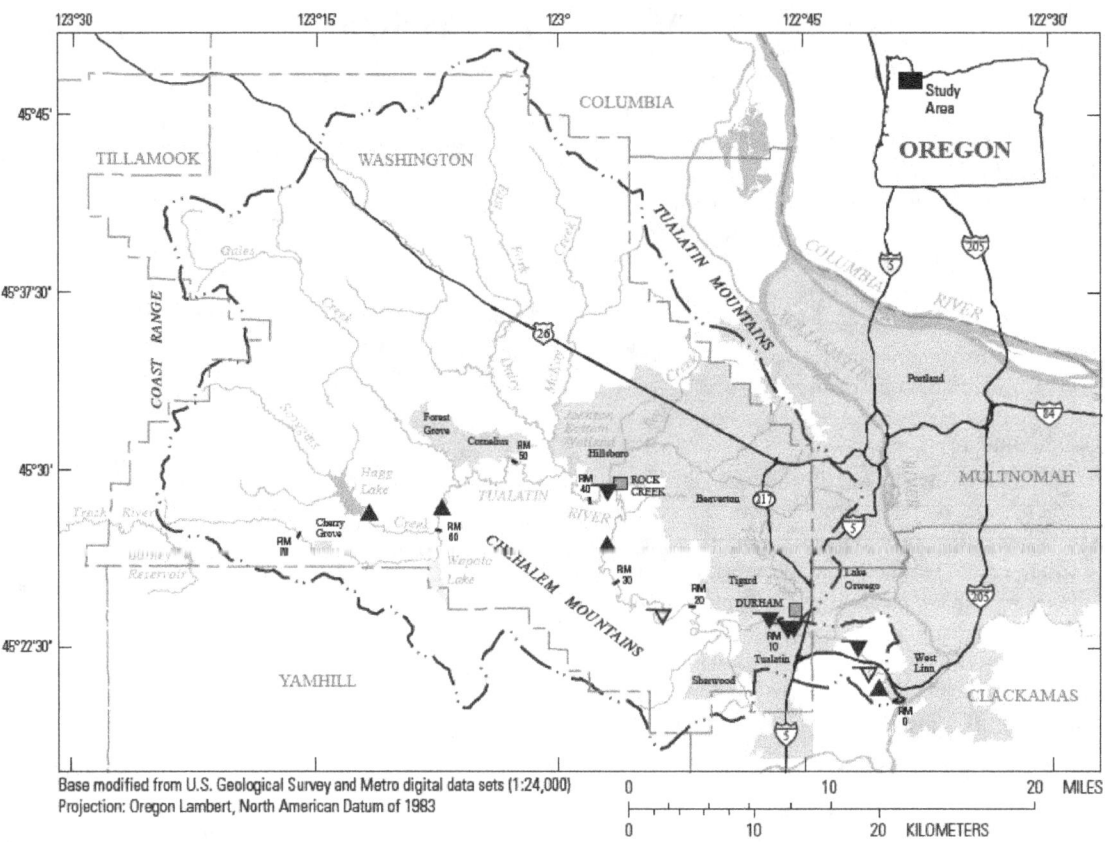

Base modified from U.S. Geological Survey and Metro digital data sets (1:24,000)
Projection: Oregon Lambert, North American Datum of 1983

EXPLANATION

▢	Designated urban growth boundary (2009)
— ·· —	Basin boundary
▣	Wastewater treatment facility
▼	Plankton sampling site
▽	Plankton sampling and continuous water-quality monitoring site
▲	Streamflow gage
RM 20	River mile

Figure 1.   Tualatin River basin, Oregon, showing the location of selected sampling sites, streamflow gages, and continuous water-quality monitors.

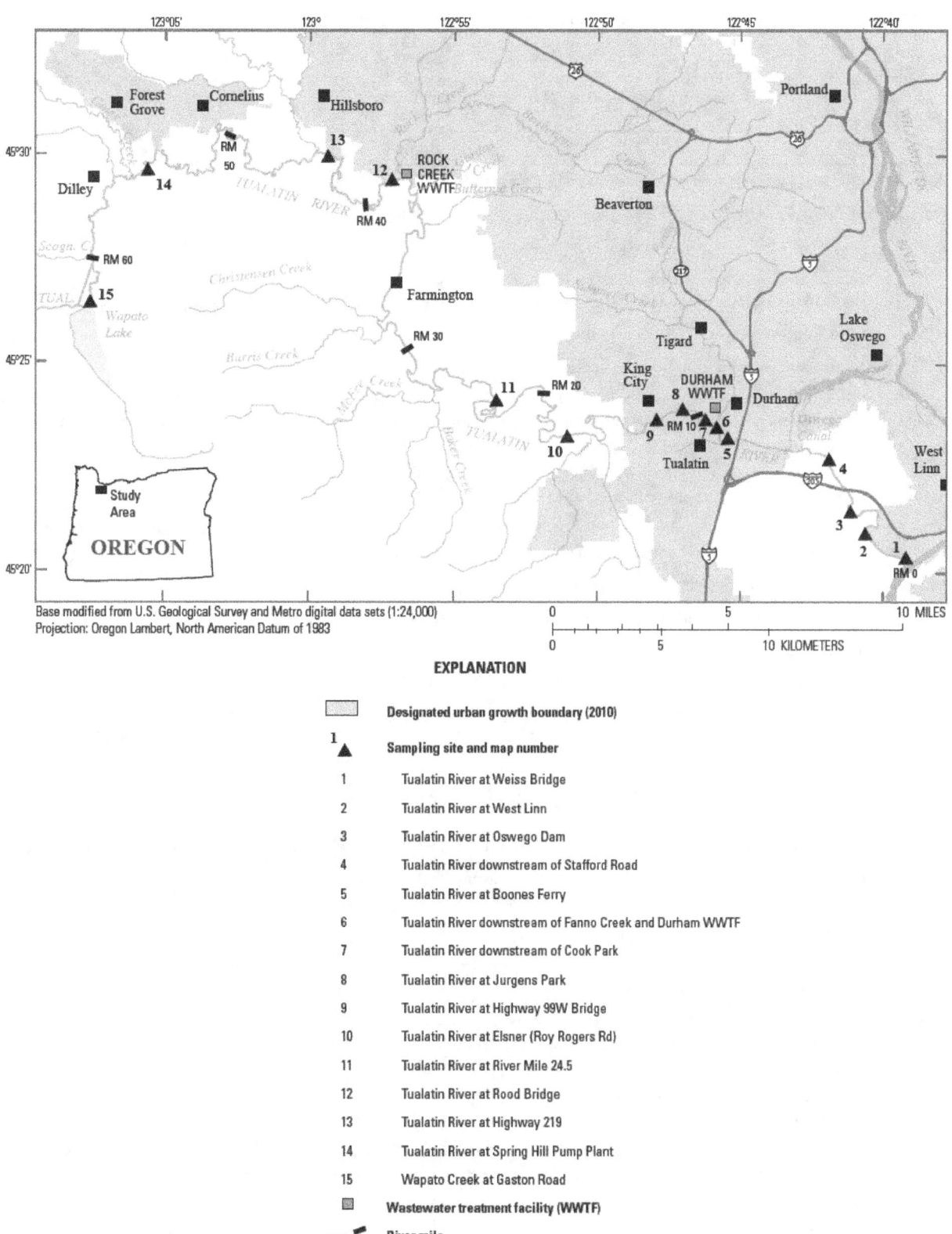

Base modified from U.S. Geological Survey and Metro digital data sets (1:24,000)
Projection: Oregon Lambert, North American Datum of 1983

EXPLANATION

Designated urban growth boundary (2010)

Sampling site and map number

1	Tualatin River at Weiss Bridge
2	Tualatin River at West Linn
3	Tualatin River at Oswego Dam
4	Tualatin River downstream of Stafford Road
5	Tualatin River at Boones Ferry
6	Tualatin River downstream of Fanno Creek and Durham WWTF
7	Tualatin River downstream of Cook Park
8	Tualatin River at Jurgens Park
9	Tualatin River at Highway 99W Bridge
10	Tualatin River at Elsner (Roy Rogers Rd)
11	Tualatin River at River Mile 24.5
12	Tualatin River at Rood Bridge
13	Tualatin River at Highway 219
14	Tualatin River at Spring Hill Pump Plant
15	Wapato Creek at Gaston Road

Wastewater treatment facility (WWTF)

RM 10   River mile

Figure 2. Tualatin River basin, Oregon, showing the location of water-quality and plankton sampling sites.

Table 1.   History of water-quality conditions and management actions in the Tualatin River basin, Oregon.

[**References:** Carter and others (1976); Oregon Department of Environmental Quality and Unified Sewerage Agency (1982). **Abbreviations:** ODEQ, Oregon Department of Environmental Quality; TMDL, Total Maximum Daily Load; USA, Unified Sewerage Agency; USGS, U.S. Geological Survey; BOD, biochemical oxygen demand; WWTF, wastewater treatment facility; RM, river mile; mL, milliliter]

Time period	Description
1960s	Raw waste in river; river often blanketed with floating algae during summer.
1966	Melosira dominant with toxic blue-green algae (*Microcystis*) during low flow.
1969	Countywide ban on new construction in urban areas until new plan developed for wastewater treatment.
1970	Unified Sewerage Agency formed to consolidate and manage wastewater treatment plants. Eldon Mills Dam (Barney Reservoir) completed on Middle Fork of North Fork Trask River with releases to Tualatin River for drinking water.
1972	Clean Water Act enacted. ODEQ lists Tualatin River as water-quality limited [303(d) listed] for aesthetics and swimming.
1972–73	Nutrient enriched, but not organically polluted; 51 genera of pollution-tolerant algae.
1975	Scoggins Dam completed; summertime releases from Henry Hagg Lake begin. Improvements in temperature, conductance, and BOD, but not in other parameters (Oregon Department of Environmental Quality and Unified Sewerage Agency, 1982).
1976	Durham Advanced Wastewater Treatment Facility comes on line replacing 14 small treatment plants. Decreased carbonaceous biochemical oxygen demand, total suspended sediment, total phosphorus, Wetter and cooler year; nutrient enriched, but not organically polluted; 31 genera of organic pollution-tolerant algae (*Stephanodiscus/Melosira* dominated).
1978	Rock Creek Advanced Wastewater Treatment Facility comes on line replacing six small treatment plants. Decreased carbonaceous biochemical oxygen demand, total suspended sediment, total phosphorus.
1977–79	Blue-green algae dominant, *Aphanizomenon flos-aquae* was major species present.
1970–80	Overall, Tualatin River showed improvements in water quality despite a 40 percent increase in population.
1980	Severe *Aphanizomenon* bloom.
1984	Tualatin River listed as water-quality limited [303(d) list] for low dissolved oxygen and nuisance algal blooms.
1986	Tualatin River listed as water-quality limited [303(d) list] for low dissolved oxygen and nuisance algal blooms. Northwest Environmental Defense Center filed suit against the U.S. Environmental Protection Agency to require that TMDLs be established for the Tualatin River.
1987	Unified Sewerage Agency (now Clean Water Services) started actively managing their share of releases from Henry Hagg Lake.
1988	ODEQ established TMDLs for ammonia-nitrogen and total phosphorus for the Tualatin River and its largest tributaries.
1990	Clean Water Services becomes the stormwater management utility for the urbanized portion of Washington County; USGS and Clean Water Services begin long-term collaboration on monitoring and research.
1991–93	USGS conducts surveys of phytoplankton, zooplankton, and water quality.
1994	All major upgrades completed at Rock Creek and Durham WWTFs to remove ammonia and phosphorus (some completed earlier); largest pH problems solved.
1998	First releases of flow augmentation water from newly expanded Barney Reservoir (expanded from 4,000 to 20,000 acre-feet). Tualatin River listed for temperature and bacteria; tributaries for same plus dissolved oxygen.
2001	ODEQ issues a revised Tualatin Basin TMDL, adding bacteria and temperature. Dissolved oxygen TMDL issued for tributaries. *Anabaena flos-aquae* and *A. planktonica* detected in Hagg Lake in June and July, peaking in late August (348 colonies per mL).
2003	In July, a loss of algal growth resulted in lower than normal dissolved oxygen levels when conditions for growing algae appeared favorable. First year Clean Water Services released water from Hagg Lake specifically for temperature trading.
2005	*Aphanizomenon* and *Anabaena* present during July–September.
2006	Diatoms and cryptomonads dominate. Low abundance of *Anabaena* in June near Scholls, RM 24.5.
2007	Bloom of the diatom *Stephanodiscus binderanus* and *Cryptomonas erosa* in the lower river in July.
2008	Oregon Department of Human Services issues a human health advisory for a bloom of *Anabaena flos-aquae* associated with the draining of the Wapato Lake agricultural area in July.

Before 1975, the Tualatin River basin had no water storage reservoirs, and the mean flow for July and August during 1929–74 averaged just 45 cubic feet per second (ft^3/s) at the West Linn stream gage (RM 1.8; figs. 1 and 2) during the dry summer period. This provided phytoplankton a considerable amount of time (possibly several weeks) to develop in the lower river. Completion of Henry Hagg Lake on Scoggins Creek in 1975 substantially increased flows in the Tualatin River, resulting in a July–August mean of about 200 ft^3/s for 1975–2010 (fig. 3).

In 1987, Clean Water Services (CWS) began actively managing releases from Hagg Lake (12,618 acre-feet (acre-ft)) to improve water-quality conditions during the critical summer low-flow periods. In 1998, releases from an expansion of Barney Reservoir (fig. 1), located in the adjacent Trask River watershed west of the Tualatin River basin, started contributing flow to the Tualatin River for water-quality augmentation and as a source of municipal drinking water. Barney Reservoir releases to the Tualatin River currently range from 0 to 40 ft^3/s during summer depending on demand and the need for flow augmentation.

Although flow augmentation has been an effective management tool for diluting some sources of nutrients, reducing water temperatures, and lessening the severity of algal blooms (CH2M Hill, 1992; Unified Sewerage Agency, 1992), this practice may hamper healthy phytoplankton development. The higher flows decrease the time available for phytoplankton to develop and have the potential to abruptly purge the system, effectively "flushing out" the algae (Rounds and others, 1999). Further, the large reductions in discharges

of ammonia and phosphorus from the two largest wastewater treatment facilities (WWTFs) in the early 1990s greatly reduced the occurrence and severity of algal blooms in the river. Ironically, although these improvements have reduced the size of algal populations and associated problems with high pH, the same factors can have deleterious implications for DO concentrations in the river as figure 4 illustrates. DO concentrations closely track phytoplankton Chl-a such that when algal populations decline, the DO can decrease to undesirable levels, sometimes for extended periods of time. The correlation between Chl-a and DO (fig. 5) also confirms the importance of photosynthesis in maintaining minimum DO levels in the river, substantiating previous findings that some moderate level of algal photosynthesis is beneficial and necessary for the river to overcome the steady loss of DO from sediment oxygen demand (SOD) created as aerobic bacteria decompose organic-rich bottom sediments (Rounds and Doyle, 1997; Rounds and others, 1999).

Simultaneously balancing the desire to prevent high phytoplankton biomass and harmful algal blooms with the need to maintain minimum DO concentrations in the river, however, is challenging. Given the importance of algal photosynthesis in maintaining a healthy DO level for the river, it is critical to understand the factors that control the size and, possibly, the composition of phytoplankton populations, especially in the lower river. That information will enable the designated management agencies to better understand how to balance these competing goals most effectively and plan for water-quality improvements.

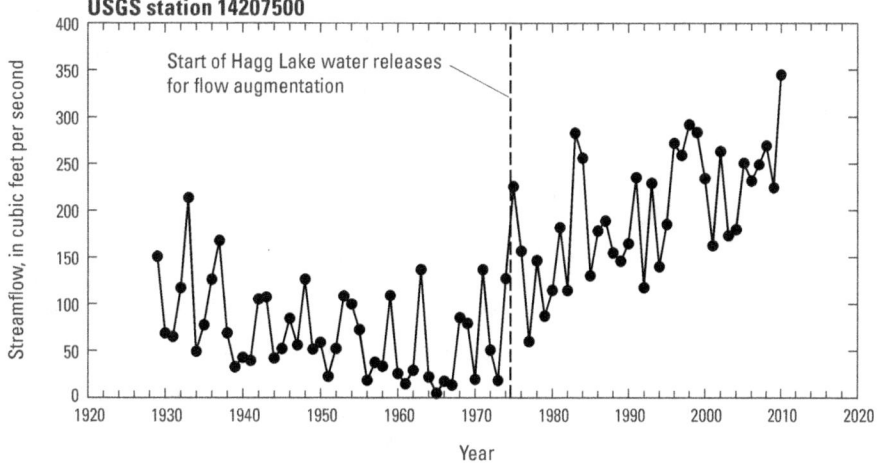

Figure 3.   Average flow in the Tualatin River at West Linn (river mile [RM] 1.8), Oregon, July and August 1929–2010.

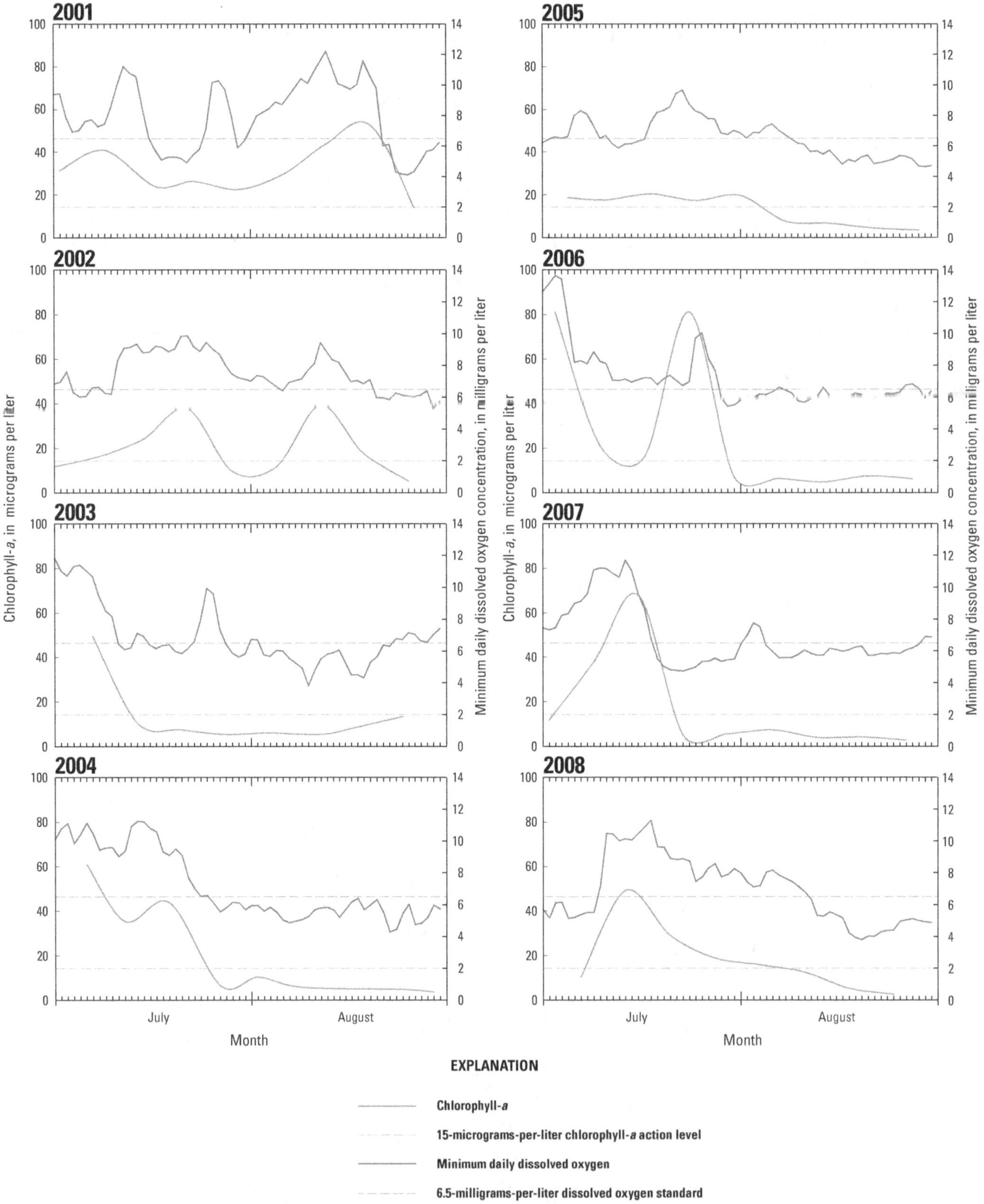

Figure 4.   Time series of phytoplankton chlorophyll-*a* and dissolved oxygen concentrations in the lower Tualatin River, Oregon, July and August 2001–08. For reference, the 6.5 milligrams per liter dissolved oxygen standard is shown, but the standard is calculated as the 30-day mean concentration, not the daily minimum values as shown. Chlorophyll-*a* for Stafford Road, RM 5.5, provided by Clean Water Services. Dissolved oxygen data from U.S. Geological Survey continuous monitor at Oswego Dam, RM 3.4.

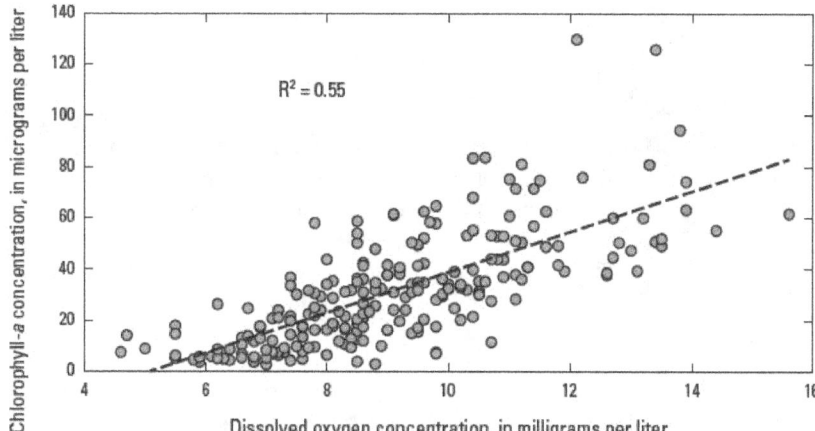

Figure 5. Chlorophyll-*a* and dissolved oxygen concentrations in the Tualatin River at Stafford Road (river mile 5.5), Oregon, July and August, 1991–2009. R^2, correlation coefficient. Data from Clean Water Services ambient monitoring program.

Patterns in Longitudinal and Seasonal Phytoplankton Abundance

Although phytoplankton populations in the Tualatin River exhibit some general longitudinal and seasonal patterns, much year-to-year variation occurs in the timing and size of blooms owing to streamflow and other factors that affect growing conditions. In the 1990s, when sampling for algae included sites in both the upper and lower river, Chl-*a* concentrations typically increased from an average of about 1 µg/L at Cherry Grove at RM 68, upstream of Scoggins Creek where Hagg Lake water enters the Tualatin River, to 3.5–4.5 µg/L in the reach between Golf Course Road and Rood Bridge, RMs 51.4 to 38.4 (fig. 6). At Farmington (RM 33.3), average Chl-*a*

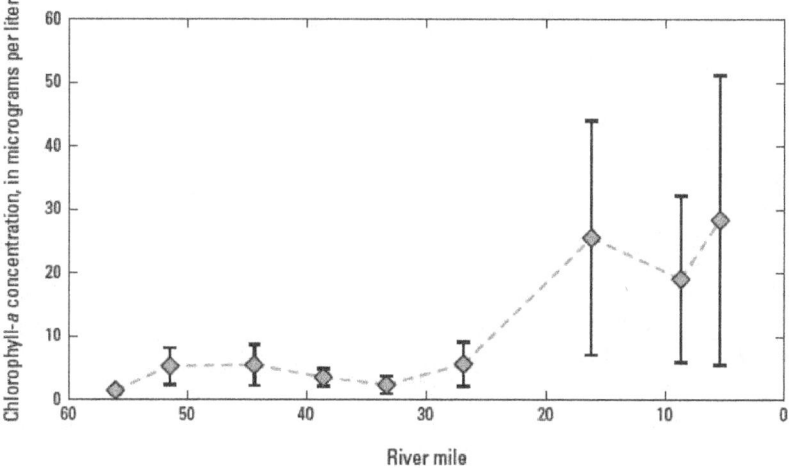

Figure 6. Longitudinal pattern in average chlorophyll-*a* concentrations in the Tualatin River, Oregon, May–October 1991–2009. Average chlorophyll-*a* concentration ±1 standard deviation for May–October. Data from the atypical 2001 (drought) and 2008 (Wapato event) years were not included. Data from Clean Water Services.

concentrations decreased to about 2.5 µg/L during summer because of dilution by discharges from the Rock Creek WWTF, which enters the river at RM 38.1 (fig. 2). The Chl-*a* concentrations increased in the reach between Farmington and Scholls, RMs 33.3 to 26.9, and continued to increase downstream to Elsner Bridge, RM 16.2, where the largest biomass values often occur. Downstream of Elsner and beyond, algal biomass is sometimes higher, but highly variable. At times, Chl-*a* concentrations increase steadily to the Oswego Dam at RM 3.4 (see photograph 1 [plankton tow samples] p. 10), while at other times, biomass levels decline, sometimes abruptly, at one or more of these sites during summer.

The average phytoplankton Chl-*a* concentrations in the Tualatin River for the June to August peak algae growing seasons, 1991–2009, are shown in figure 7. This plot excludes samples from the atypical 2001 and 2008 years, when severe drought (2001) and an unusual algal bloom resulting from Wapato Lake discharges (2008; see section, "Wapato Lake Algal Bloom") occurred. A seasonal increase in phytoplankton biomass occurs in spring; from April to May, Chl-*a* levels increased to over 10 µg/L at Elsner, RM 16.2 (fig. 7). Previous studies of Tualatin River phytoplankton in 1991–93 (Rounds and others, 1999) found the highest growth rates to occur in June or early July, lower rates in mid-summer, and slightly higher rates in late August or September, based on carbon uptake experiments. While growth rates show this trend, algal Chl-*a* typically peaks in July, when streamflows are lower, days are long, light is abundant, and water temperatures are warm. Chl-*a* levels then decline in August and September in response to shorter days, lower light availability, higher flows, or other factors.

Lower Chl-*a* concentrations in July and August are clearly evident in data from 2002 and later, denoted with dashed lines in figure 7. This decline in algal abundance is one of the reasons for conducting this study and is discussed in more detail later in the report.

38.4   24.5   10.8   9.5   9.2   5.4   3.4

Main-stem river mile

Photograph 1: Plankton tow sample vials showing longitudinal changes in plankton abundance and composition. Photograph by Kurt Carpenter, June 6, 2006.

Primary Factors Controlling Phytoplankton Growth

Previous USGS studies determined the primary factors controlling phytoplankton populations and water-quality conditions in the Tualatin River. Water-quality models calibrated for 1991–93 (Rounds and others, 1999) and 1991–97 (Rounds and Wood, 2001) indicated that streamflow, light conditions, water temperature, and phosphorus concentrations were most influential in determining the onset, duration, and magnitude of phytoplankton blooms in the Tualatin River. With these four variables, the

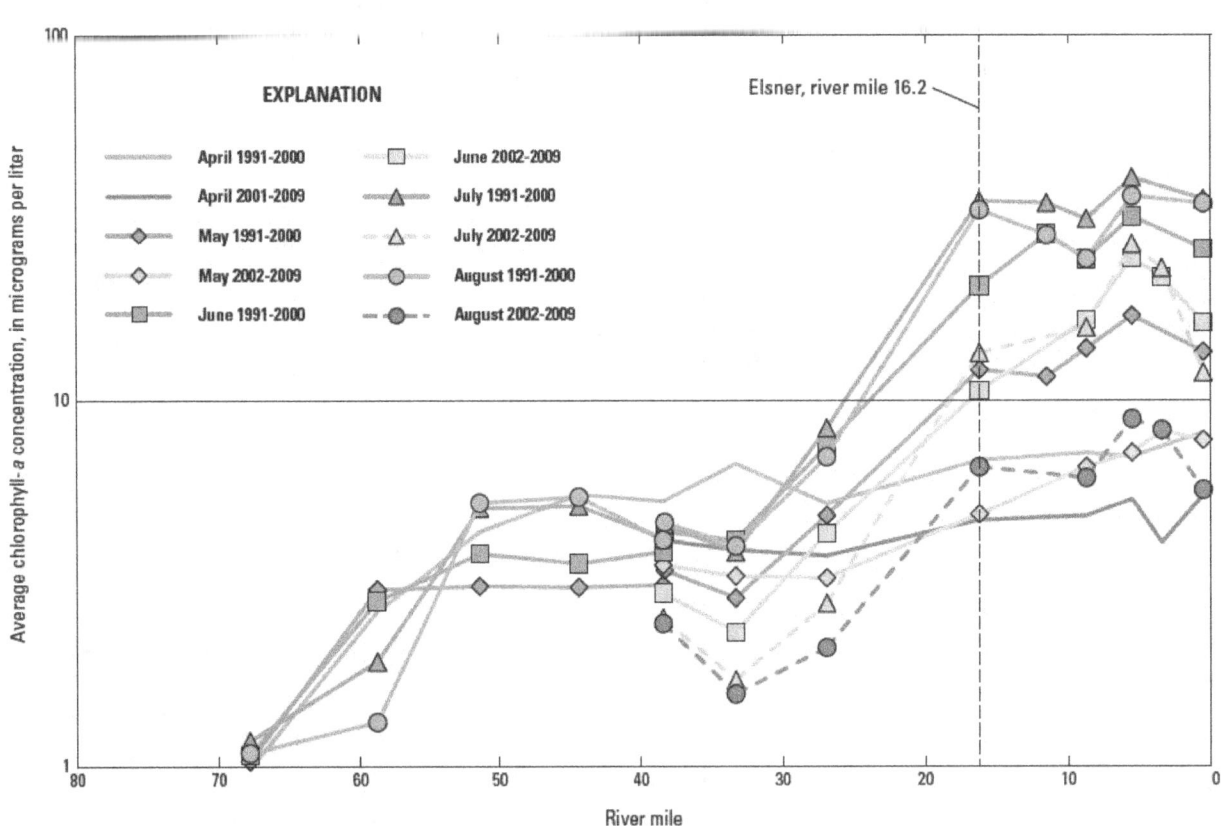

Figure 7.   Changes in the average monthly phytoplankton chlorophyll-*a* in the Tualatin River, Oregon, highlighting the 1991–2000 and 2002–2009 periods. Note y-axis log scale. Data from the atypical 2001 (drought) and 2008 (Wapato event) years were not included. Data provided by Clean Water Services.

models can simulate, with acceptable accuracy, the onset, duration, and general magnitude of phytoplankton blooms. While zooplankton grazing of algal cells occasionally was found to be an important loss mechanism for phytoplankton populations, it was limited to the lowermost reaches of the Tualatin River and only occurred during certain periods (Rounds and others, 1999).

Although the USGS Tualatin River water-quality models can predict many important constituents much of the time, errors in the simulated DO and SRP concentrations are largely due to inaccuracies in the size of the modeled phytoplankton population, at times accounting for 50 percent of the errors (Rounds and Wood, 2001). This uncertainty in the modeled phytoplankton populations arises, in part, from the models' oversimplification of the population to just one algal type. In addition, the large degree of annual variability in algal populations (figs. 4 and 6) and lack of a consistent successional pattern in the algal species composition (Doyle and Caldwell, 1996) has made it difficult to incorporate interspecies differences in buoyancy, nutrient requirements, growth rates, or susceptibility to grazing by zooplankton, for example, into the models.

Study Background, Objectives, Scope, and Approach

During the 1990s, phytoplankton populations in the Tualatin River would typically grow throughout the summer and decline when flows increased or light decreased such that algal growth was no longer favorable. These conditions occurred most often during the later part of the growing season (late August or early September). Algal populations in the lower river were mostly sustained above 25 µg Chl-a/L in July and August and DO concentrations generally stayed above the State DO standard for the river—a three-level tiered standard—based on (1) the 30-day mean concentration of 6.5 mg/L, (2) the 7-day average of the daily minimum concentration of 5 mg/L, and (3) the absolute minimum concentration of 4 mg/L. While episodes of lower DO would occasionally occur when blooms crashed or algal populations subsided, this typically occurred at the end of the growing season, in late summer or early fall (Jan Miller, retired Clean Water Services Water Resources Program Manager, oral commun., 2010). Beginning in 2002 or 2003, however, Chl-a levels have been notably lower, particularly in August, and the frequency of periods when the DO has not met one or more of the State standard has increased (fig. 8).

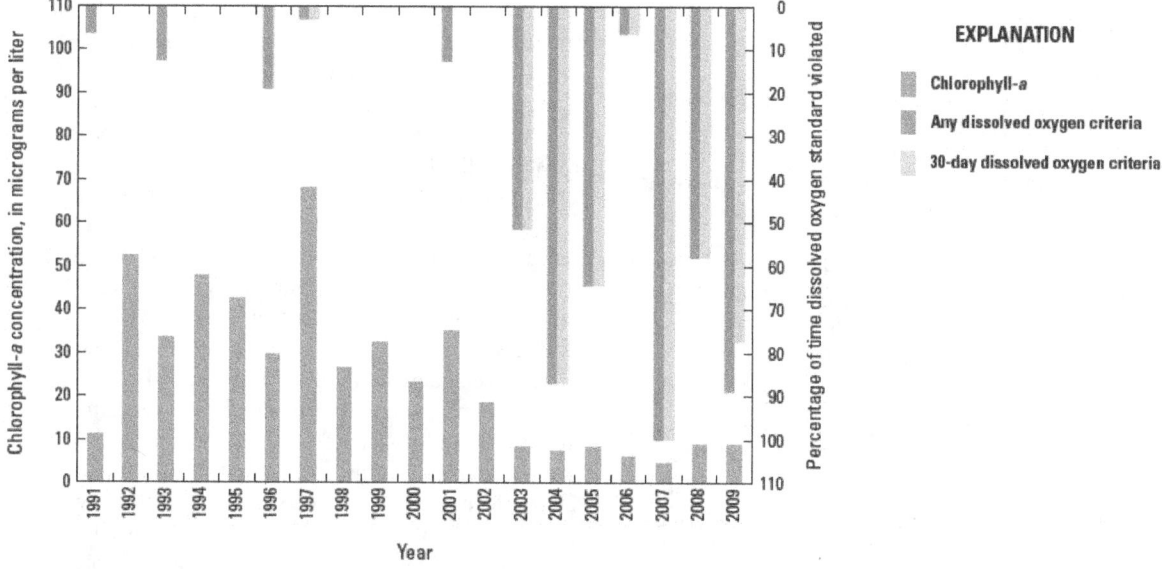

Figure 8. Trends in the average August chlorophyll-a concentration and the percentage of time in August that the minimum dissolved oxygen (DO) standard was violated in the lower Tualatin River, Oregon, 1991–2009. Chlorophyll data are from the Stafford Road site (river mile 5.5); DO data are from the Oswego Dam site (RM 3.4). The dark blue bars represent days when any of the three DO criteria were violated (instantaneous, 7-day minimum mean, 30-day mean); the light blue bars are for days when only the 30-day mean DO criteria was violated. The 6.5-milligram per liter State standard is based on the 30-day mean concentration.

Since about 2003, the simple model of residence time, light, temperature, and phosphorus has been insufficient to explain the absence of even moderate algal populations in the river for some periods in July and August. Phytoplankton populations in recent years often are smaller or terminate earlier than in years past, despite apparently favorable flow, light, and nutrient conditions. To exacerbate matters, the large decline in phytoplankton abundance now often coincides with the seasonal peak in water temperature in late July or early August, which because of the lower DO solubility at higher temperatures, can lower DO concentrations even further.

In response to these declines in phytoplankton abundance and episodes of low DO, the USGS partnered with CWS to investigate the relative importance of the factors influencing phytoplankton and formulate and evaluate hypotheses to explain the causes of such declines. Because DO is such an important indicator of river health, and because DO is needed to offset oxygen demands in the river, it is critical that the causes of such reductions in phytoplankton abundance be identified.

The study objectives were to characterize the phytoplankton populations during the 2006–08 growing seasons; examine how phytoplankton respond to factors such as streamflow, sources of flow, weather, zooplankton, and selected water-quality parameters; and identify those factors contributing to the recent declines in phytoplankton populations. Some of the questions that this study was designed to answer included:

1. What is the current composition of the phytoplankton community in the Tualatin River during summer?

2. What are the most important environmental variables influencing phytoplankton species composition?

3. Why does phytoplankton abundance decline abruptly (typically in July or early August) and not rebound when streamflow, sunlight, and other conditions otherwise appear to be favorable?

To address these questions, both new and existing data on plankton and water quality were analyzed. New plankton and water-quality data were collected in the Tualatin River at 2- to 3-week intervals during the growing season, typically May–September for 3 years. Longitudinal sampling of the river extended from either the Rood Bridge site (RM 38.4) or the sampling station near Scholls (RM 24.5) to the Oswego Dam (RM 3.4) at a half-dozen sites per trip on 5–6 occasions each year (table 2, fig. 2). Other data used in the analyses included streamflow and water-quality data from continuous monitors, and results from discrete water-quality samples from the CWS routine watershed monitoring program. In certain instances, the scope was expanded beyond the intensive 2006–08 time period to include more data from 1991 to 2009

to frame the larger issues. In particular, ten bloom/crash cycles during 2003–08 were examined in detail to help identify specific causal factors. The resulting dataset was large and complex, and multivariate and other statistical techniques were required to weigh the importance of the various environmental factors influencing phytoplankton assemblages. In 2008, in-stream bioassay experiments were conducted to test the effect of WWTF effluent on phytoplankton biomass (Chl-a) and DO production, and phosphorus additions were used to examine the potential for phosphorus to limit algal growth.

Hydrology and Study Area Description

The hydrology of the Tualatin River and surrounding landscapes have been modified through water withdrawals for municipal, agricultural, and other uses, as well as discharges from tile drains, reservoir releases, treated wastewater, and stormwater runoff. Each modification has had an effect on water quality. This section provides a brief overview of the basin and the important hydrologic conditions that affect phytoplankton populations; more information is available from Rounds and others (1999) and other reports listed in the section, "References Cited."

The Tualatin River can be divided into four reaches based on stream gradient and geomorphology (Rounds and others, 1999; Bonn, 2008), and the different characteristics of these reaches are critical in influencing the abundance of phytoplankton. In the headwater reach upstream of RM 55, the river is narrow and relatively steep, particularly in the Coast Range. The Spring Hill Pump Plant withdraws water in this reach to supply the Tualatin Valley Irrigation District (TVID) as well as the Joint Water Commission (JWC), which provides drinking water to many cities in the western part of the basin.

Downstream from RM 55, the Tualatin River meanders through a 25-mi section where the average gradient decreases to just 1.3 ft/mi., then beginning around RM 30, the river takes on a reservoir-like character that extends to the Oswego Dam at RM 3.4 (see photographs 2–5, p. 14). This "reservoir reach" has an average gradient of 0.1 ft/mi, is wide enough (>150 ft) to minimize shading from streamside vegetation, averages about 10–15 ft of water depth, and includes a few atypical pools that reach maximum depths of 25–30 ft (Rounds, 2002). This reach is deep enough and has a high enough concentration of colloidal and suspended particles, that light does not reach the river bottom except at the margins (Bonn and Rounds, 2010). As a result, algal production is dominated by phytoplankton that float with the water and grow mainly in the top 5–10 ft of the water column; the river can be thought of as a conveyor belt that provides a certain amount of time for algae to grow before transporting them downstream, to be replaced by new populations from upstream—provided there is ample inocula.

Table 2.   Sampling sites and data collection activities in the Tualatin River basin, Oregon, 2006–08.

[**Bolded** sites were sampled for the majority of the longitudinal plankton surveys. For a more complete listing of river mile (RM) indexes, refer to Bonn (2008). **Abbreviations:** USGS, U.S. Geological Survey; WWTF, wastewater treatment facility]

Location	Main-stem river mile	Plankton sampling	Water-quality sampling	Continuous water-quality monitor	Streamflow gage	Inflow	Outflow
Tualatin River at Weiss Bridge	0.2		X				
Tualatin River at West Linn	1.8				X		
Tualatin River at the Oswego Dam	3.4	X	X	X			
Tualatin River downstream of Stafford Road Bridge	[1]5.4/5.5	X	X				
Tualatin River at the Oswego Canal headgates	6.7						X
Tualatin River at Boones Ferry	8.7		X				
Tualatin River downstream of Fanno Creek and Durham WWTF	**9.2**	X	X				
Tualatin River at Durham WWTF Release / Fanno Creek	9.3					X	
Tualatin River downstream of Cook Park	**9.7**	X	X				
Tualatin River at RM 9.9 near Tualatin	9.9			X			
Tualatin River at RM 10.2	10.2	X					
Tualatin River at RM 10.6	10.6	X					
Tualatin River at Jurgens Park	**10.8**	X	X				
Tualatin River downstream from the Highway 99W Bridge	11.1	X					
Tualatin River at Highway 99W Bridge	11.6	X	X				
Tualatin River at RM 13.5	13.5	X					
Tualatin River at Elsner (Roy Rogers Rd)	16.2		X				
Tualatin River at RM 17.3	17.3	X					
Tualatin River at RM 17.9	17.9	X					
Tualatin River at RM 20.4	20.4	X					
Tualatin River at RM 21.1	21.1	X					
Tualatin River at RM 23.2	23.2	X					
Tualatin River at River Mile 24.5	**24.5**	X	X	X			
Tualatin River at Scholls Bridge	26.9		X				
Tualatin River at Farmington	33.3		X		X		
Tualatin River at Rock Creek WWTF Release	38.1					X	
Tualatin River at Rood Bridge	**38.4**	X	X		X		
Tualatin River at Golf Course Road	51.5		X				
Tualatin River at JWC/TVID Spring Hill Pump Plant Intake	56.1						X
Tualatin River at Dilley	58.8				X		
Scoggins Creek Below Henry Hagg Lake	—[2]			X	X		
Scoggins Reservoir (Hagg Lake) Release	—[3]					X	
Tualatin River at Wapato Improvement District Return Flow	60.1		X			X	
Tualatin River at Wapato Improvement District Headgate	61.9						X
Tualatin River at Cherry Grove	67.8		X				
Tualatin River at City of Hillsboro Haine's Falls Intake	73.3						X
Barney Reservoir Aqueduct Outfall	78					X	

[1] USGS plankton site is located 0.1 mile downstream from the bridge where Clean Water Services samples.

[2] Scoggins Creek site is located 0.1 mile downstream from Hagg Lake.

[3] Hagg Lake outfall is located at RM 5.1 on Scoggins Creek.

Photograph 2: Tualatin River at the water-quality monitoring station near Scholls (RM 24.5). Photograph by Kurt Carpenter, July 19, 2007.

Photograph 3: Tualatin River at the Highway 99W Bridge put-in (RM 11.6). Photograph by Kurt Carpenter, July 19, 2007.

Photograph 4: Tualatin River downstream of Fanno Creek (RM 9.2). Photograph by Kurt Carpenter, August 9, 2005.

Photograph 5: Tualatin River at the Oswego Dam (RM 3.4). Photograph by Kurt Carpenter, August 9, 2005.

A growth curve of increasing phytoplankton populations with downstream distance in the reservoir reach (as in fig. 6) may be shifted downstream if flows increase, and variations in algal biomass and DO production occur in response to the growing conditions, primarily driven in summer by the amount of sunshine and magnitude of flow. Because of these spatial and temporal effects, algal populations at any given location are governed by the particular suite of upstream conditions integrated over several days to a couple of weeks or longer. Events that increase flow are especially influential as they shorten the time available for algae to grow before they are transported out of the system downstream.

Residence times in the reservoir reach during summer have historically been long enough, as much as 10–14 days or longer, for phytoplankton populations to develop into substantial blooms. Previous studies (Rounds and others, 1999) found that flows less than about 300 ft^3/s were required to allow enough residence time for phytoplankton populations to become substantial at the Oswego Dam water-quality monitoring site (fig. 2). Additional factors affecting residence time in the reservoir reach included large water withdrawals (50 ft^3/s) from the Tualatin River at the Oswego Canal (RM 6.7) through 1995, as well as the use of plywood flashboards on the concrete weir of the Oswego Dam to raise water levels in the river 1–2 ft; this facilitated water withdrawals at the Oswego Canal via gravity into Lake Oswego (fig. 2). Oswego Canal withdrawals decreased substantially after 1995 but continued to average about 10–15 ft^3/s through 2003, after which the withdrawal rate decreased to 1 ft^3/s or less. The use of flashboards was common until the late 1990s when fewer boards were placed on the weir. Variable use of the flashboards continued through about 2005, but their use was discontinued thereafter.

Beyond the Oswego Dam, from RM 3.4 to the Tualatin River's mouth, is the lowermost "riffle reach" of the Tualatin River. The river gradient in this reach increases to 13 ft/mi and the river passes through a narrow gorge before joining the Willamette River south of Portland. Monitoring sites in this reach include the USGS streamflow gaging station at West Linn (RM 1.8) and the CWS water-quality sampling station at Weiss Bridge (RM 0.5).

Sample Collection and Data Analysis Methods

Water-Sample Collection and Processing

Water samples were collected by USGS for analyses of total and dissolved nutrients, Chl-*a*, and selected major ions and metals with clean 2-liter (L) polyethylene bottles using clean techniques, and samples stored on ice prior to laboratory analysis. Samples collected during the USGS longitudinal surveys in 2006–08 were grab samples taken from just beneath the water surface, typically at the center of the stream cross section, although at two stations, the Tualatin River at RM 24.5 and at Jurgens Park (RM 10.8), samples were collected from a shore-based floating dock, when a canoe was not available. The Oswego Dam site (RM 3.4) was sampled near the fish ladder along the left bank.

Water samples collected by CWS were grab-integrated-composite (GRIC) samples collected at five points across the river width, and integrated over the depth of the river or as much as 10 feet, whichever was less, and five subsamples were composited in a churn splitter. Water samples were processed and analyzed by the CWS Water-Quality Laboratory in Hillsboro, Oregon, using methods published in their watershed monitoring plan (Clean Water Services, 2006). Water samples for dissolved constituents were passed through 0.45-micron syringe filters prior to analysis, and subsamples for Chl-*a* were collected onto glass fiber filters and analyzed fluorometrically (Clean Water Services, 2006).

Chl-*a* samples for the WWTF effluent experiments were analyzed at the USGS Oregon Water Science Center (ORWSC) in Portland, Oregon, using a Turner Designs fluorometer. Filters were ground in 90 percent acetone and fluorescence values were obtained before and after the addition of acid to account for the presence of phaeopigments according to methods described by the American Public Health Association (1992).

Field Parameters and Streamflow

Instantaneous field parameters (water temperature, DO, pH, specific conductance, turbidity, and Chl-*a*) were measured with a Yellow Springs Instruments (YSI) multiparameter sonde during USGS longitudinal surveys, and these discrete measurements complimented continuous data from nearby water-quality monitors located at RM 24.5, Cook Park (RM 9.9), and Oswego Dam (RM 3.4) (fig. 2). Continuous streamflow data were obtained from an Oregon Water Resources Department (OWRD) gage located in the Tualatin River at Farmington (RM 33.3) and the USGS gage at West Linn (RM 1.8). Flow data for Scoggins Creek downstream of Henry Hagg Lake were obtained from the Bureau of Reclamation. Releases from Barney Reservoir and Hagg Lake for flow augmentation were obtained from the Tualatin River Flow Management Technical Committee (Bonn, 2006; 2007; 2008). In this report, natural flow is defined as any flow not derived from upstream reservoirs or WWTFs, and flow augmentation water is flow derived from both Barney Reservoir and Hagg Lake minus all known water withdrawals.

Plankton-Sample Collection, Processing, and Identifications

Phytoplankton grab samples were collected from just below the river surface into 250-mL polyethylene bottles and placed on ice. Zooplankton samples were collected with a 12-in. diameter, 80-µm mesh plankton net which was hand tossed from a canoe or from shore, capturing approximately 10 ft of towing distance per sample for a total volume of approximately 222 L, or 0.222 cubic meter (m^3) per sample. The sample was washed down the net toward the "cod" end with repeated rinses, into a 20-mL plastic vial. Phytoplankton samples were preserved with 2.5 mL of Lugol's solution (1 percent final concentration), and zooplankton samples were preserved with isopropyl alcohol (25 percent final concentration, by volume). Additional tows of "net plankton" were collected at each site for microscopic observation of the unpreserved plankton community.

Preserved phytoplankton samples were shipped to Aquatic Analysts, Inc., in Friday Harbor, Washington, for identification and enumeration. Permanent microscope slides were prepared for each sample by filtering an appropriate aliquot of the sample through a 0.45-µm membrane filter (APHA Standard Methods, 1992). A section of filter was cut out and placed on a glass slide with immersion oil added to make the filter transparent. A cover slip was placed on top, with nail polish applied to the periphery for permanency. Most algae were identified by cross-referencing several taxonomic sources. A minimum of 100 algal units, defined as discrete particles (either cells, colonies, or filaments), were counted along a measured transect on a microscope slide with a Zeiss standard microscope using 1000X magnification. Only algae with intact chloroplast and believed to be alive at the time of collection were counted. Average biovolume estimates of each species were obtained from calculations of microscopic measurements of each alga type in each sample analyzed. The number of cells per colony or the length of a filament was recorded to determine the biovolume per unit-alga conversion factors, which were used to calculate the total biovolume per taxon for each sample. In this report, biovolume is used along with Chl-*a* (continuous measurement at the two water-quality monitors and discrete measurement of water samples analyzed in the laboratory) to estimate algal abundance in the water column.

Zooplankton samples were shipped to ZP's Taxonomic Services (Lakewood, Washington), for identification and enumeration. Zooplankton densities, in number of organisms per cubic meter, were determined for each sample by counting a minimum target of 400 organisms or, if fewer organisms were present, the entire sample.

Plankton Bioassay Experiments

Streamside experiments were conducted using 300-mL glass BOD bottles to examine the potential effects of WWTF effluent (0, 30, and 50 percent) and soluble reactive phosphorus (SRP) additions of 20 and 50 µg/L on Chl-a and DO production during incubations lasting 6–24 hours.

Water for the experiments was collected from the Tualatin River at Rood Bridge (RM 38.4, upstream of the Rock Creek WWTF discharge location), and transferred to an 18-L churn splitter. WWTF effluent was collected from the Rock Creek treatment facility just prior to its discharge to the river. Varying amounts of river water and WWTF effluent were dispensed into BOD bottles with a graduated cylinder to obtain a range of effluent concentrations, either 0 and 50 percent, or 0, 30, and 50 percent, depending on the experiment. The starting conditions in each bottle were either measured (for DO) or estimated from measurements of Chl-a and nutrient concentrations in the water sources used to prepare the initial samples (unamended native water, plankton-amended water, and treated effluent), taking into account the proportions of each source. Ending conditions were determined by directly analyzing the contents of each sample bottle.

On the morning of each experiment, one large 20-L carboy was filled with native water collected from the Tualatin River at Rood Bridge and one 10-L carboy of treated effluent was obtained. The filled carboys were suspended in the river to attain temperature equilibrium, and then they were removed to a shady streamside location where the native water carboy was purged for about a minute with gaseous N_2 to reduce the DO concentration to approximately 75–95 percent of saturation. Purging with N_2 did not alter the pH appreciably (by removing CO_2), and gave the phytoplankton some "room" to photosynthesize. Despite this effort, production of excessive DO sometimes occurred within the bottles, forming oxygen bubbles in the supersaturated conditions. Because this gas (DO) tends to escape from the BOD bottle when the stopper is removed, some of the DO measurements likely underestimated actual DO production.

"Plankton-amended" water was created by adding several 20-mL vials of concentrated plankton collected from the Tualatin River at Jurgens Park (RM 10.8) using an 80-µm mesh net to river water collected from the Rood Bridge site. The plankton-amended samples contained 0.3–1.9 times the amount of phytoplankton biovolume compared with algal

biovolume in the Tualatin River at Jurgens Park on the day of the experiment. Because samples were collected with a plankton net, however, the zooplankton densities in the amended samples were much higher—containing as much as 24 times the amount of zooplankton. During the July 15–17, 2008 experiment, the zooplankton density attained in the amended sample bottles was within the range of zooplankton abundance at some of the downstream sites, and was more representative of actual conditions compared with earlier experiments.

Experiments were set up such that a range of WWTF effluent and phosphorus concentrations were tested. To test for potential phosphorus limitation, some bottles were spiked with a concentrated solution of KH_2PO_4 using a micropipette to boost SRP concentrations by 20 µg/L (low dose) or 50 µg/L (high dose).

Bottles were topped off with a small volume of river water from Rood Bridge to allow proper removal of air bubbles and stoppered. Once all bottles were filled, the DO and water temperature were measured using a Yellow Springs, Inc. Clark-type DO probe equipped with an electric stirrer. The bottles were randomly arranged in two metal wire baskets placed in the river on the upstream side of the Oswego Dam next to the fish ladder, in an exposed location that receives afternoon sunshine. Bottles were incubated at a depth of about one foot in an area with sufficient flow to ensure that bottles were maintained at river temperature.

Following incubation, the DO in each bottle was measured on site, bottles were placed on ice in a cooler, and samples were transported to the laboratory for further processing. Chl-a samples were collected on 47-mm diameter 0.7-µm GF/F glass fiber filters and frozen prior to analysis at the USGS ORWSC. Samples for dissolved nutrients—nitrite-plus-nitrate ($NO_2^- + NO_3^-$), ammonia-plus-ammonium ($NH_3 + NH_4^+$), hereafter referred to as ammonia, and SRP—were filtered through 25-mm diameter 0.45-µm Acrodisk™ filters using a 60-mL syringe, refrigerated, and then transported to the CWS Water-Quality Laboratory in Hillsboro, Oregon, for analysis. In this report, all constituent concentrations are given in atomic units (as N or P, for example).

Quality Assurance Data

All of the CWS data used in this study, including nutrients, Chl-a, turbidity, and field parameters (water temperature, pH, DO, specific conductance), were subjected to a comprehensive and rigorous quality assurance (QA) procedure by the CWS Water-Quality Laboratory. CWS laboratory methods and protocols have been reviewed by the USGS Branch of Quality Systems and were determined to be suitable. Field methods in use by CWS, including the collection of samples using depth- and width-integrating techniques and the use of churn splitters for subsampling, are consistent with USGS procedures. In addition, the CWS laboratory has been certified by the National Environmental Laboratory Accreditation Conference (NELAC), participates in a monthly-to-quarterly QA program with the USGS ORWSC for nutrients and Chl-*a*, and participates in many laboratory performance tests such as the twice-a-year USGS Standard Reference Sample (SRS) program, a national inter-laboratory comparison study (U.S. Geological Survey, 2013). SRS results from many years of participation in the program have shown that the CWS laboratory consistently produces high quality data that are sufficiently accurate for all the parameters used in this study.

Results from quality-assurance samples collected specifically for this study are shown in appendix A (tables A-1 to A-3), and the results are summarized below. Replicate water-quality samples collected during this study (table A-1) revealed no major problems. Replicate samples for Chl-*a*, nitrite-plus-nitrate, and SRP had percent relative differences (PRD) of less than 5 percent. The PRD for replicate dissolved ammonia analyses was higher (28 percent), but at these low concentrations (0.014 and 0.018 mg/L), such variability in replicates is not uncommon and does not affect the interpretations in this report.

Replicate samples for phytoplankton identification and enumeration showed some variations in species composition and identifications (table A-2), but for the most part, the dominant taxa were the same or similar, and overall, the reproducibility was deemed acceptable. More variability was found in the zooplankton counts (table A-3), and there was a greater amount of variability in the replicate zooplankton tow-net samples compared with the phytoplankton samples. This may have been due to the greater heterogeneity of zooplankton in surface waters, to larger spatial scales—a 10-ft transect compared with grab samples of surface water—or longer amount of time (and thus more movement across the water) for repeat zooplankton sample collections compared with phytoplankton. In any case, the reproducibility in the zooplankton taxa identifications was acceptable for the dominant taxa.

Data Analysis Methods

Plankton, flow, water-quality, and meteorological data (solar radiation and rainfall) were analyzed using a combination of standard regression and multivariate statistical methods. Longitudinal, time-series, and other plots were generated to examine patterns in the data. Spearman rank correlation, Principal Components Analysis (PCA), and a variety of multivariate statistical analyses were performed using the computer software package PRIMER (Plymouth Routines In Multivariate Ecological Research), version 6 (Clarke and Gorley, 2006). Environmental data were log transformed and standardized (or "normalized" in PRIMER) prior to their use in the multivariate analyses.

Species composition data were analyzed using nonmetric multidimensional scaling (NMDS) ordinations and other methods in PRIMER including BEST, ANOSIM, and SIMPER. NMDS ordination analyses were used to understand patterns in the phytoplankton species data, and to identify potential subsets of samples for further analysis. Bio-Env+Stepwise (BEST) analyses were performed to test the relative importance of various factors in explaining variations or patterns in the phytoplankton species composition. The strength of each variable, or combination of variables, is represented by the rho value and associated P value. PRIMER also tests the statistical significance of the selected combination of variables (global rho (R) statistic) or individual rho value using a Monte-Carlo permutation simulation. Analysis of similarity (ANOSIM), the nonparametric equivalent to analysis of variance (ANOVA), was used to test for significant differences among selected pre-defined groups of samples, such as those collected on different dates or during different flow or phytoplankton biomass (Chl-*a*) condition. Then the similarity percentages (SIMPER) procedure was used to calculate the average species abundance in each group to identify algal taxa that might explain any differences.

Many of the water-quality and flow variables were autocorrelated, with suites of variables explaining similar sources of variation within the data. Chloride, for example, was highly correlated with specific conductance, nitrate, and percent flow from WWTF, owing to the relatively high conductance and nitrate concentrations in WWTF effluent. In addition, because the percentages for each flow source (WWTFs, reservoir augmentation, and natural flow) are not independent of each other, they often were correlated. In these cases, one "best" representative or "surrogate" variable was selected for the final BEST analysis, based on the strength of its individual correlation with the phytoplankton assemblage data. The removal of redundant variables did not reduce the explanatory power of the analysis substantially, and had the benefit of simplifying the interpretation. Through an iterative process, unimportant and redundant variables were removed until a final solution of 1–5 variables was obtained. In cases where removing a flow variable from the model resulted in a much lower overall correlation, multiple flow variables were permitted in the final solution.

NMDS ordinations of phytoplankton samples were constructed from Bray-Curtis similarity matrices based on square-root transformed algal biovolume data. The NMDS ordination routines in PRIMER work iteratively to optimize a solution whereby samples having higher similarity are plotted close together and samples with lower similarity are plotted farther apart. The underlying similarity matrix used to construct the ordinations was correlated with each of the environmental variables to identify those variables with the most significant correlations (rho values).

In several cases, water chemistry data from CWS were used to fill gaps in the data set. For this purpose, CWS data from the Tualatin River at Scholls Bridge (RM 26.9) were used for the USGS sampling site at RM 24.5; data from Elsner (RM 16.2) were used for the Highway 99W/Jurgens Park sites (RMs 11.6/10.8); data from Boones Ferry (RM 8.7) were used for the site downstream of Fanno Creek (RM 9.2); and in 2006 only, data from Weiss Bridge (RM 0.5) were used to fill gaps at the Oswego Dam site (RM 3.4). In most cases, the substitute data were collected by CWS on the same day, and always within 2–3 days of the plankton sample collections.

Several flow variables were included in the final multivariate environmental dataset, including total streamflow, natural streamflow (total and percent of streamflow without WWTF inputs or flow augmentation), flow augmentation level (total and percent of reservoir releases minus any withdrawals), and WWTF effluent flow (total and percent). Flow variables were calculated for each sampling site and adjusted to the nearest day to account for (a) travel time between a streamflow gage and the sampling sites, and (b) the distance between inputs of water from Barney Reservoir, Hagg Lake, or inputs of treated effluent from the Rock Creek and Durham WWTFs, to the downstream sampling sites. Time lags for streamflow measurements also were applied to calculate streamflow for each of the sampling sites on the day of sampling. Effluent discharge data were obtained from Bonn (2006, 2007, 2008). Streamflow data from the Tualatin River at Farmington gage were obtained from OWRD and used to calculate the flow and effluent percentages for samples collected at sites from Rood Bridge to Cook Park (RMs 38.4–9.7). Total streamflow, flow augmentation, and WWTF effluent percentages for sites downstream of Cook Park, which are affected by inputs from the Durham WWTF and Fanno Creek at RM 9.3, were calculated using streamflow data from the USGS gage at West Linn (RM 1.8).

Travel times for the 2006–08 "summer" period (May–September) were estimated from a number of sources, including measured and modeled travel times published by Rounds and others (1999), from results of dye tracer studies conducted by USGS on September 14–15, 1992 (Lee, 1995), and from modeled results from other dye tracer studies. Water velocities and estimated travel times were assigned according to whether sampling occurred during three categories of flow conditions (190–220, 240–275, and 285–310 ft^3/s; table 3). The flow ranges were determined by evaluating break points in the distribution of flow data at West Linn for summer months (June–September) for 2006–08. These streamflow ranges correspond to the majority of the flow conditions during the USGS sampling, although higher flows did occur in May 2008: 567 ft^3/s at West Linn. The timing of flow augmentation from Barney Reservoir, located to the west in the Trask River basin (fig. 1), was estimated by adding 2 days of travel time to the estimated travel time from Hagg Lake to each sampling site (Bonn, 2008).

Table 3. Estimated average water velocities used to estimate travel times in four reaches of the Tualatin River, Oregon, over a range of summer streamflows.

Reach (length)	Streamflow in the Tualatin River at West Linn (river mile 1.8) (cubic feet per second)		
	190–220	240–275	285–310
	Water velocity (miles per hour)		
Hagg Lake to Rood Bridge (26.6 miles)	1	1	1
Rood Bridge to RM 24.5 (13.9 miles)	0.30	0.40	0.50
RM 24.5 to Elsner (8.3 miles)	0.14	0.18	0.22
Elsner to Oswego Dam (12.8 miles)	0.10	0.12	0.14
Oswego Dam to Weiss Bridge (2.9 miles)	0.70	0.85	1.00

Climate, Streamflow, and Water-Quality Conditions

To put the results from this 2006–08 study into perspective, selected climatological, streamflow, Chl-a, and other water-quality variables were compared to a longer historical record going back to 1991 (fig. 9). Although conditions were variable among the 3 study years, they were not atypical for many basic parameters, including rainfall, solar radiation, water temperature, or specific conductance. The nitrate levels for 2006–08, however, were among the highest during 1991–2009.

Streamflow and Sources of Flow

The magnitude and source of streamflow can affect phytoplankton through mechanisms including water chemistry, temperature, travel time, and other factors. Changes in the source of flow may alter the amount or composition of plankton inoculum entering the river, thereby affecting downstream algal communities.

Differences in winter precipitation, and the timing of snowmelt and late-spring rains, resulted in higher late-May and early-June flows in 2006 and 2008 compared to 2007, but by late June or early July, flows were similar among the 3 years. Compared with the 1991–2010 average flow at West Linn of 215 ft^3/s in July and August, flows of 215, 245, and 261 ft^3/s in 2006–08 were average or somewhat above average.

Summertime flow in the Tualatin River originates from four sources: (1) natural flow from the headwaters, wetlands, tributaries, and groundwater; (2) flow augmentation from two upstream reservoirs; (3) discharges from the two largest WWTFs (Rock Creek and Durham); and (4) discharges and return flows from agricultural areas and managed wetlands in the basin. These sources maintained streamflow levels

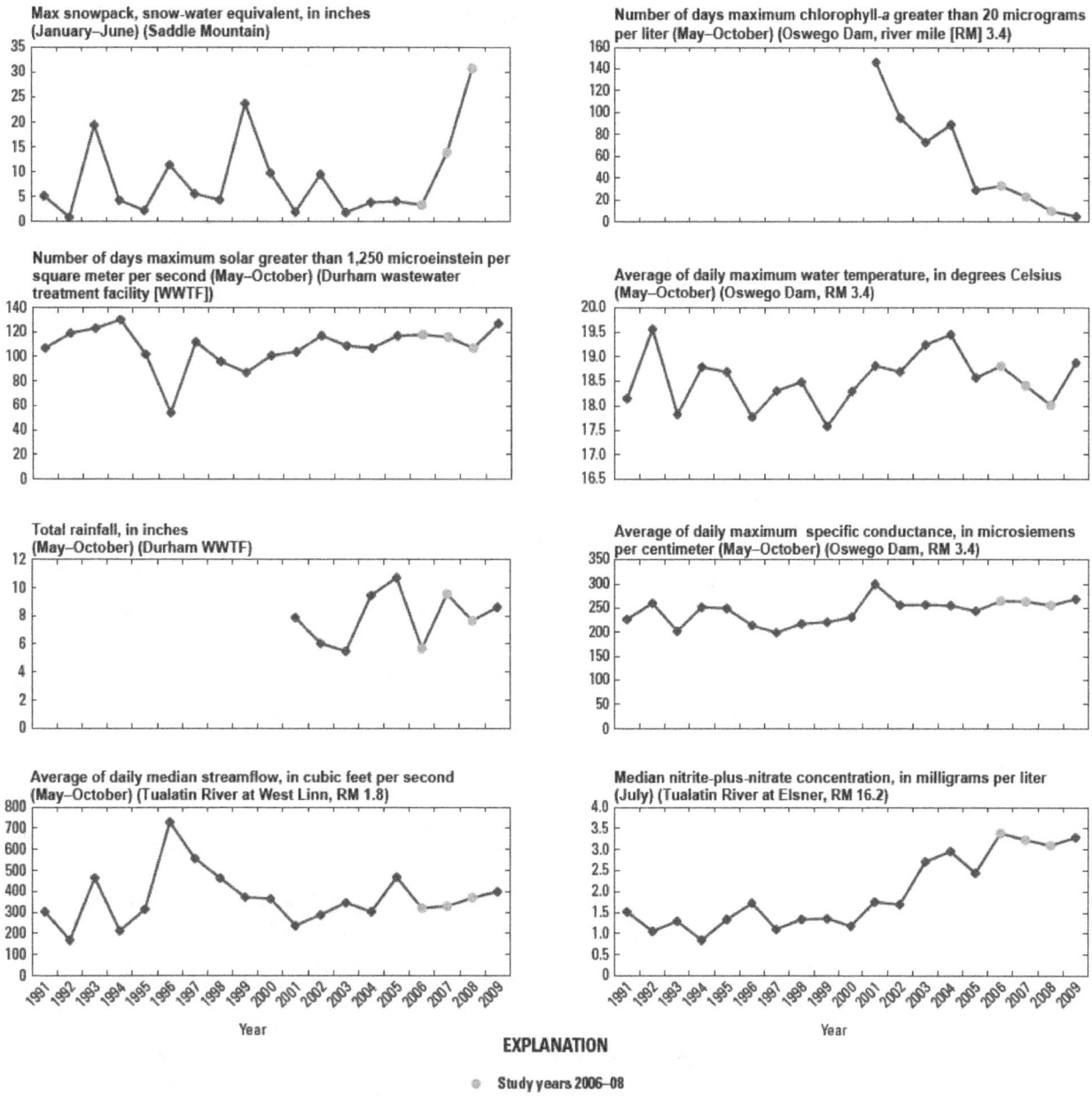

Figure 9. Annual patterns in snowpack, solar radiation, streamflow, water-quality, and algal conditions in the Tualatin River, Oregon, 1991–2009.

near or above 200 ft³/s at West Linn during 2006–08. Flow measurements from various locations and knowledge of travel time through the river system is routinely used to estimate proportions of these flow sources at key locations to understand how they influence water-quality conditions (Bonn, 2006; 2007; 2008; see fig. 10).

Natural flow accounted for 70–80 percent of the total flow at Farmington and West Linn in early summer (June) in 2006–08, declining to as low as 38–40 percent later in summer as reservoir releases for irrigation, water supply, and flow augmentation, and treated WWTF-effluent percentages increased (fig. 11).

Patterns in flow augmentation were similar during 2006–08, with most releases beginning in earnest in late June and accounting for roughly 20–30 percent of the streamflow at West Linn in late summer. The releases were somewhat more consistent in 2006, in contrast to 2007 and 2008 when short periods of reduced releases from reservoirs occurred because rainfall decreased the need for flow augmentation (fig. 11). Note that flow augmentation was calculated as reservoir releases minus withdrawals from the Cherry Grove, Patton, Wapato, TVID Spring Hill, and JWC Spring Hill intakes, and sometimes not all of the reported withdrawals are made, so flow augmentation values may, at times, be higher.

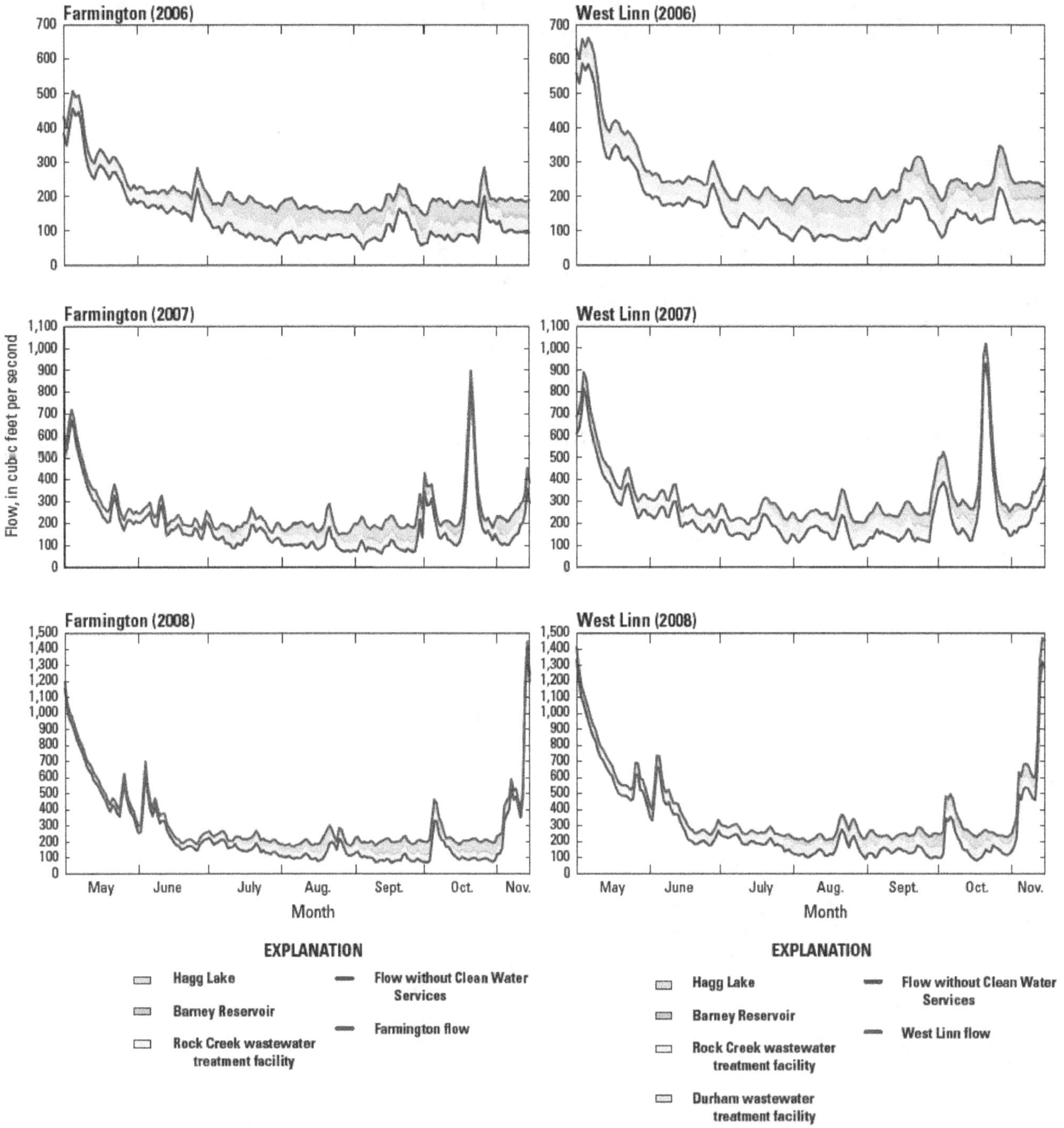

Figure 10.   Natural flow, wastewater treatment facility discharges, and flow augmentation in the Tualatin River at Farmington (river mile [RM] 33.3) and West Linn (RM 1.8), Oregon, 2006–08. Figures modified after Bonn (2006, 2007, 2008).

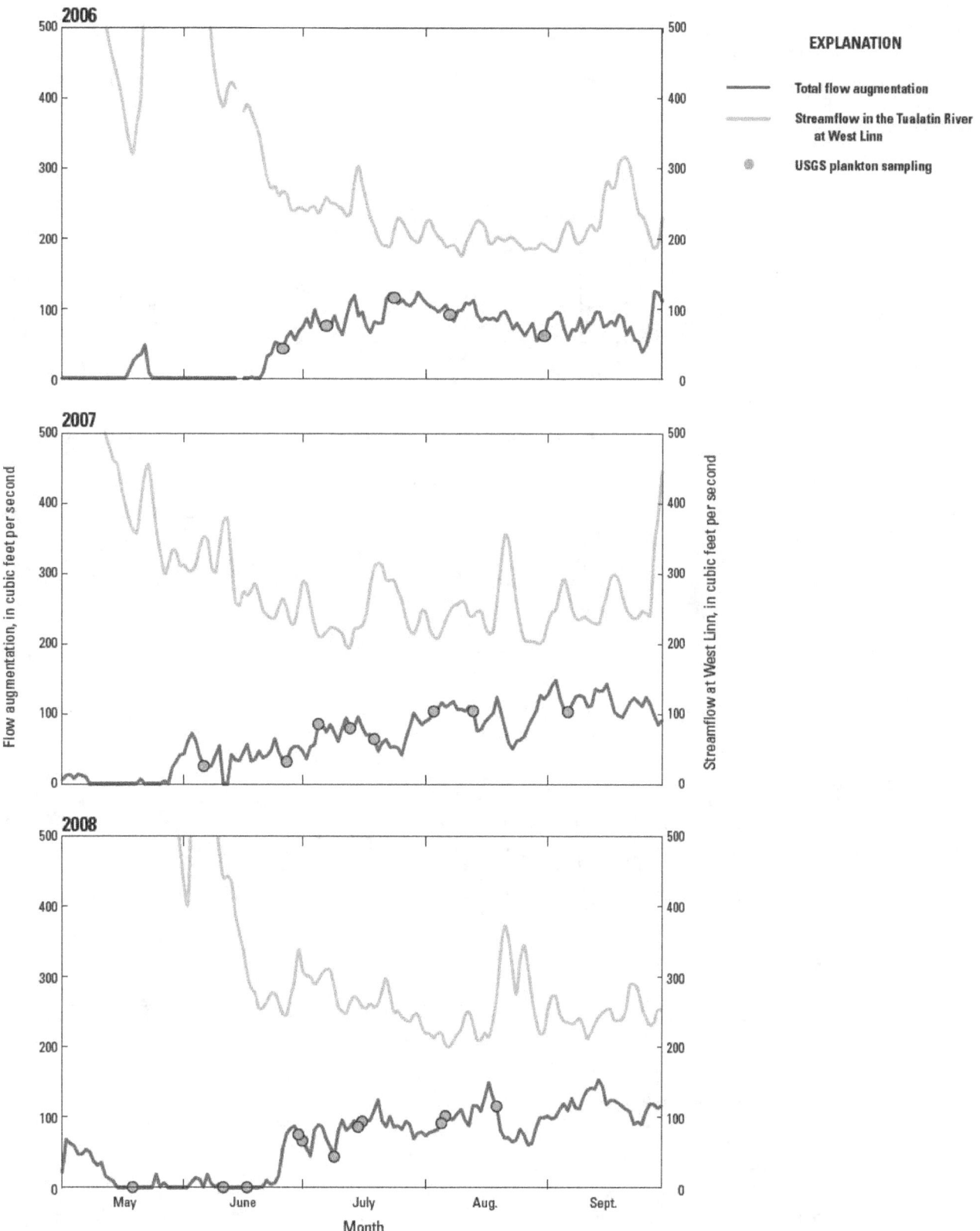

Figure 11.   Streamflow and flow augmentation in the lower Tualatin River, Oregon, showing the timing of longitudinal samplings, 2006–08. Flow augmentation includes reservoir releases minus withdrawals from Hillsboro-Cherry Grove, Patton, Wapato, TVID Spring Hill, and JWC Spring Hill intakes.

Effluent discharged to the Tualatin River from the Rock Creek WWTF contributed as much as 25–31 percent of the total streamflow at Farmington during May–October of 2006, 2007, and 2008 (fig. 12); downstream of the Durham WWTF, treated effluent accounted for as much as 33–38 percent of the total streamflow at West Linn (fig. 12). Discharges from the

WWTFs have increased over the last 20 years, particularly from the Rock Creek WWTF (fig. 13) as population growth in that part of Washington County increased. As a result, the percentage of treated effluent in the river downstream of the Rock Creek WWTF also has increased over the years.

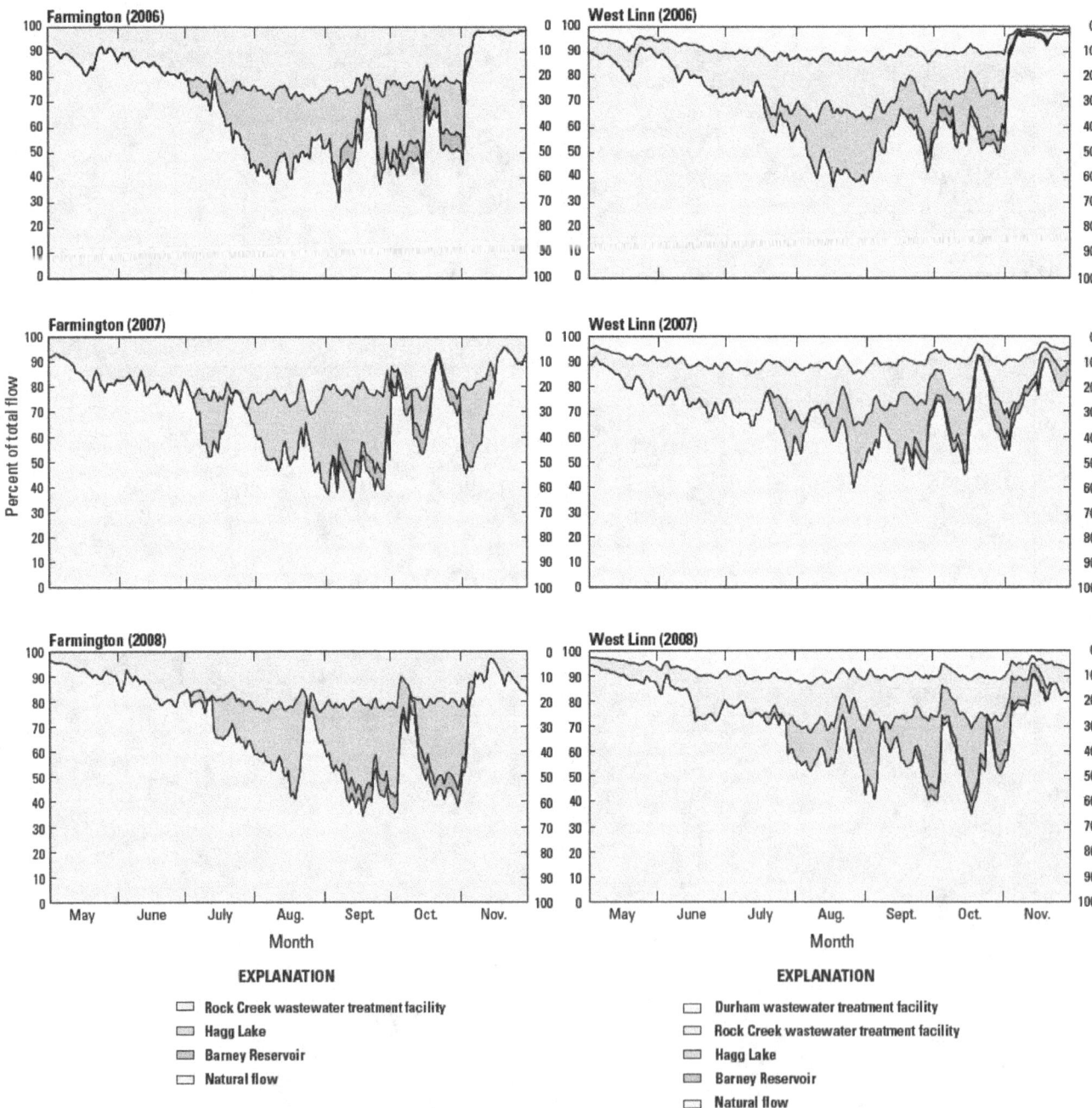

EXPLANATION

☐ Rock Creek wastewater treatment facility
▨ Hagg Lake
▨ Barney Reservoir
☐ Natural flow

EXPLANATION

☐ Durham wastewater treatment facility
☐ Rock Creek wastewater treatment facility
▨ Hagg Lake
▨ Barney Reservoir
☐ Natural flow

Figure 12.   Percentage of natural flow, flow augmentation, and wastewater treatment facility effluent in the Tualatin River at Farmington (river mile [RM] 33.3), and West Linn (RM 1.75), Oregon, 2006–08. Flow augmentation includes reservoir releases minus withdrawals from Hillsboro-Cherry Grove, Patton, Wapato, TVID Spring Hill, and JWC Spring Hill intakes.

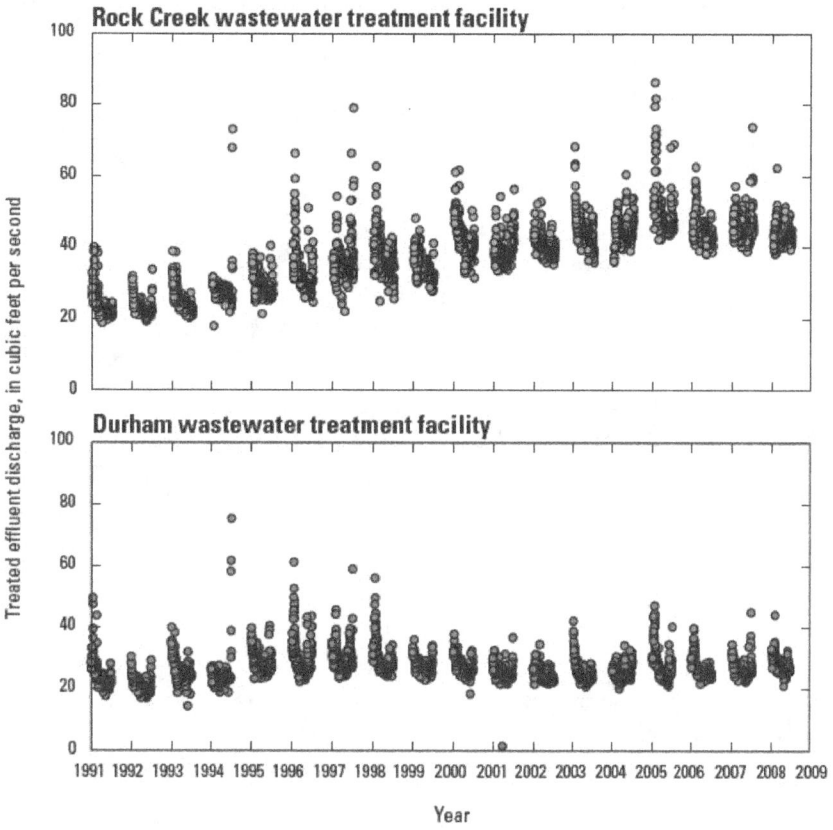

Figure 13.   Daily treated effluent discharges from the Rock Creek and Durham wastewater treatment facilities, Oregon, May–October 1991–2008. Data from Clean Water Services.

The quality of treated effluent from these two WWTFs is shown in table 4. The WWTFs convert ammonia to nitrate during summer when temperatures are warm enough to support populations of nitrifying bacteria in the treatment plants. As a result, WWTF effluent during summer is characterized by low concentrations of ammonia but high concentrations of nitrate. Phosphorus removal is required to meet TMDL limits during May through October, resulting in relatively low median SRP concentrations ranging from 0.013 to 0.018 mg/L in 2006–08 (table 4). It is not clear how much of this phosphorus is available to phytoplankton; studies in other treatment facilities that utilize alum for tertiary phosphorus removal seem to indicate that much of the phosphorus may not be bioavailable (Li and Brett, 2010).

Bioavailable Nutrients

Dissolved nutrient concentrations in the lower Tualatin River (fig. 14) were generally high enough to support algal growth during most of the year. Patterns in nutrient concentrations reflected the seasonal pattern in streamflow variations (source and amount) and patterns in algal growth. Nitrogen concentrations in the river are controlled mainly by discharges from the Rock Creek WWTF (RM 38.1). Nitrification of ammonia during treatment results in high concentrations of nitrate-nitrogen in effluent, causing concentrations to increase to 4–5 mg/L during summer at sites downstream (note the large increase between Rood Bridge and Scholls in fig. 14). Concentrations of ammonia, on the other hand, tend to be low and occasionally decline during algal blooms to less than 0.005 mg/L at Stafford Road (RM 5.5).

Table 4.   Quality of treated effluent from the Rock Creek and Durham wastewater treatment facilities, Oregon, May–October 2006–08.

[Data from Clean Water Services. **Abbreviations:** C, degrees Celsius; WWTF, wastewater treatment facility; mg/L, milligram per liter]

Parameter	Units		Rock Creek WWTF			Durham WWTF		
			2006	2007	2008	2006	2007	2008
Effluent temperature	°C	Minimum	16.7	17.8	16.7	15.7	16.5	15.4
		Median	21.3	21.3	21.3	21.2	20.3	20.2
		Maximum	23.3	23.3	23.5	23.0	22.4	23.3
Dissolved oxygen	mg/L	Minimum	7.3	7.1	7.5	7.3	6.3	7.4
		Median	8.5	8.6	8.7	8.9	8.9	9.1
		Maximum	9.7	9.8	11.1	10.3	10.2	11.7
Chloride	mg/L	Minimum	40.8	36.7	42.6	42.0	40.2	43.5
		Median	51.3	51.6	51.2	55.1	59.0	51.7
		Maximum	55.6	56.2	57.7	61.0	65.9	59.9
Ammonia	mg/L	Minimum	0.01	0.01	0.01	0.03	0.02	0.03
		Median	0.03	0.03	0.03	0.07	0.04	0.05
		Maximum	11.20	9.51	13.00	1.33	1.18	9.86
Nitrite+nitrate	mg/L	Minimum	9.5	10.6	8.9	6.8	6.0	3.7
		Median	14.5	14.4	15.4	10.2	9.7	10.0
		Maximum	18.8	18.7	19.7	12.6	14.8	14.1
Total Kjeldahl nitrogen	mg/L	Minimum	0.90	0.83	0.84	0.93	0.96	1.01
		Median	1.2	1.2	1.3	1.5	1.5	1.6
		Maximum	12.2	11.3	13.8	2.4	2.7	11.2
Soluble reactive phosphorus	mg/L	Minimum	0.005	0.005	0.005	0.005	0.005	0.005
		Median	0.018	0.014	0.013	0.016	0.016	0.013
		Maximum	0.66	0.66	0.78	0.32	0.75	0.76
Total phosphorus	mg/L	Minimum	0.025	0.025	0.034	0.025	0.025	0.035
		Median	0.072	0.060	0.078	0.088	0.080	0.092
		Maximum	0.80	0.88	1.03	0.50	0.89	1.01

Despite the fact that ammonia may be a preferred source of nitrogen for most algae, the nitrate in the river should be available for algal growth; therefore, it appears unlikely that nitrogen would limit the growth of most phytoplankton species in the river downstream of the Rock Creek WWTF.

Periods of high phytoplankton abundance often corresponded to sharp declines in ammonia and SRP concentrations. These dissolved nutrients are taken up and converted into algal biomass, which resulted in negative relations between concentrations of Chl-*a* and these nutrients each summer, particularly in 2007 (fig. 15). The ammonia and

SRP concentrations were sometimes below detection during algal-bloom periods when Chl-*a* concentrations were highest. Some unusually high concentrations of SRP were measured at Rood Bridge in July 2008 (fig. 14). Additional water testing verified the source to be drainage water discharged from the Wapato Lake agricultural area, where a levee failure in the previous winter caused this low-lying area to remain ponded until early summer (Bonn, 2008). The elevated ammonia concentration at Stafford Road in early July 2008 was due to a discharge from the Rock Creek WWTF (Jan Miller, Clean Water Services, written commun., 2011).

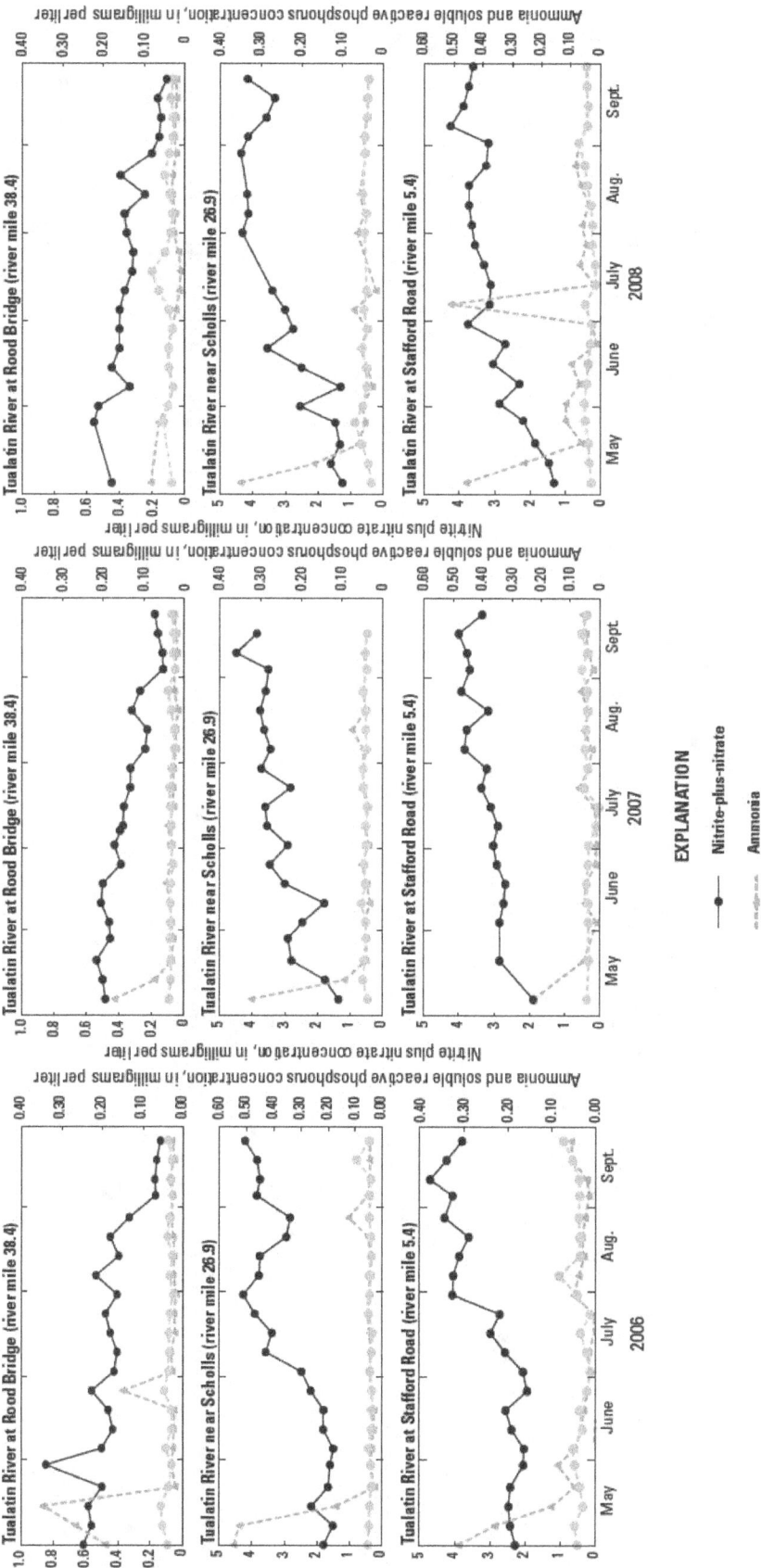

Figure 14. Time series of dissolved nitrite-plus-nitrate, ammonia, and soluble reactive phosphorus concentrations at select sites in the Tualatin River, Oregon, 2006–08. Data from Clean Water Services.

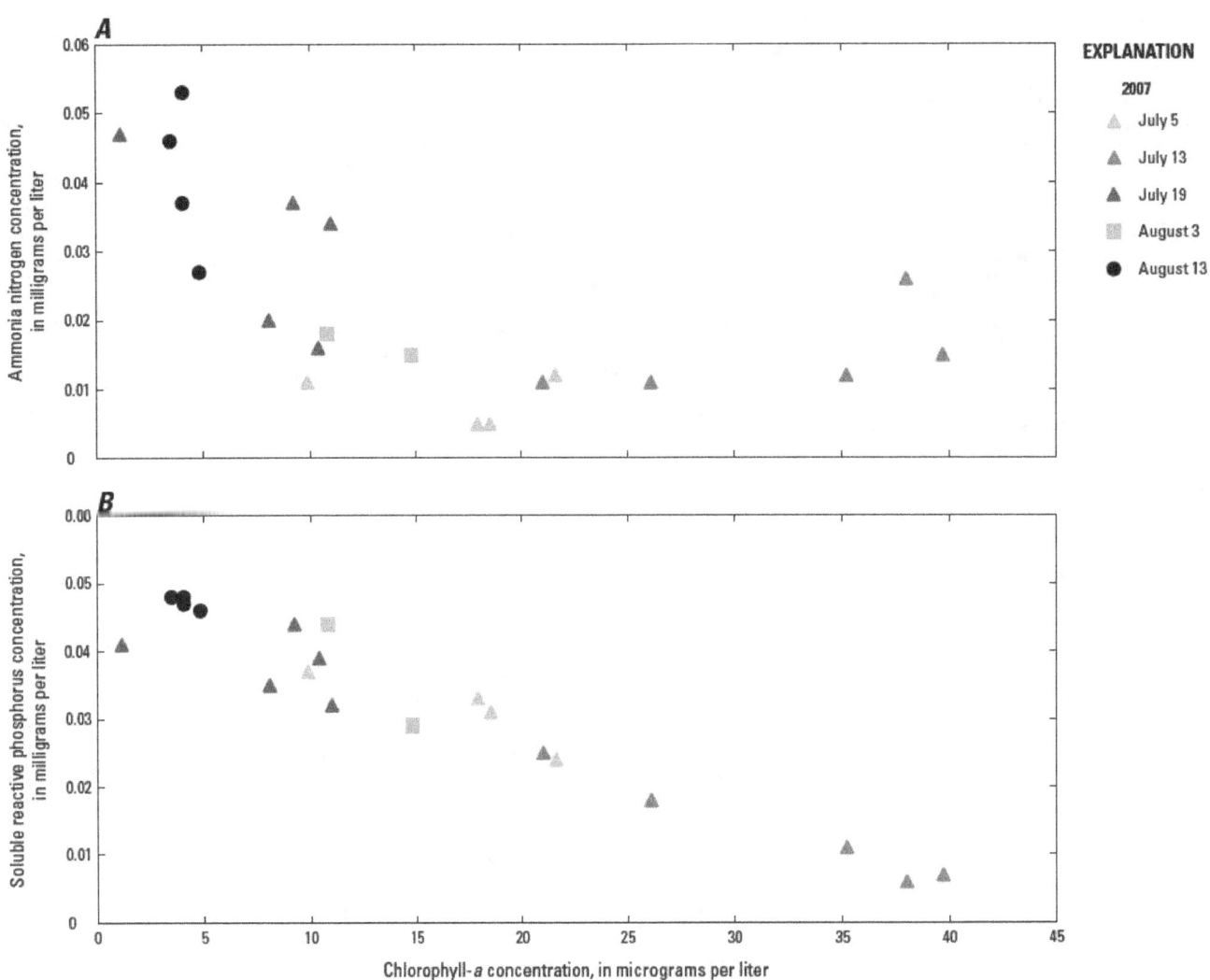

Figure 15. Chlorophyll-*a* and (*A*) ammonia and (*B*) soluble reactive phosphorus in the lower Tualatin River, Oregon, 2007. Lower Tualatin River includes the Highway 99W and Jurgen's Park sites (river miles 11.6 and 10.8, respectively) downstream to Oswego Dam (river mile 3.4).

Patterns in Plankton Populations 27

Patterns in Plankton Populations

Phytoplankton

Annual, longitudinal, and seasonal patterns in phytoplankton Chl-*a* reveal much about the cycles and dynamics of Tualatin River algal communities (figs. 4, 6, and 7). Substantial phytoplankton populations developed each year; concentrations at Stafford Road peaked well above the 15-μg/L action level (fig. 4). Chl-*a* concentrations peaked in July at 80, 70, and 50 μg/L in 2006, 2007, and 2008, respectively, and then declined to below 5–10 μg/L in late July or early August. In 2006, a June–July bloom was followed by a rapid decline, but the population then rebounded to similar concentrations only to decline again at the end of July. In 2007, a single bloom occurred, peaking in mid-July and declining a week later. In 2008, a bloom also formed in mid-July (a result of Wapato Lake discharges, discussed below), and after that the bloom declined slowly, eventually producing the lowest concentrations of DO observed during this study—less than 4 μg/L at the Oswego Dam (fig. 4).

Phytoplankton populations in the Tualatin River are composed of a variety of species, including diatoms, green and blue-green algae, and small algal flagellates. Although the algal assemblages are dynamic and diverse, just a few groups made up the majority of the total algal biovolume in the lower river during much of the growing season in 2006–08. A total of 143 algal taxa were identified in 117 main-stem samples in 2006–08 (table 5; appendix B). The dominant algal taxa during summer included filamentous centric diatoms *Stephanodiscus binderanus*, *S. hantzschii*, and several *Cyclotella* and *Aulacoseira* (formerly *Melosira*) species, small flagellated green algae, *Chlamydomonas* sp., cryptophytes including *Cryptomonas erosa* and *Rhodomonas minuta*, and occasional but sometimes large blooms of blue-green algae, mainly *Anabaena flos-aquae* and *Aphanizomenon flos-aquae* (see photographs 6–16, p. 27–28).

Table 5. Taxa richness and abundance (biovolume) of major algal groups (divisions) in the main-stem Tualatin River, Oregon, 2006–08.

[Data from 117 phytoplankton samples collected from the main-stem Tualatin River]

Algal division	Total number of taxa	Percent of total number of taxa	Total biovolume	Percent of total biovolume
Bacillariophyta (diatoms)	93	65	42,910,833	38.5
Cryptophyta (cryptophyte algae)	2	1	42,586,213	38.2
Cyanobacteria (blue-green algae)	6	4	22,039,143	19.8
Chlorophyta (green algae)	28	20	3,044,019	2.7
Chrysophyta (golden algae)	6	4	437,434	0.4
Pyrrophyta (dinoflagellates)	3	2	382,223	0.3
Euglenophyta (euglenoid algae)	3	2	67,243	0.1
Unknown classification	2	1	64,113	0.1
Total	143			

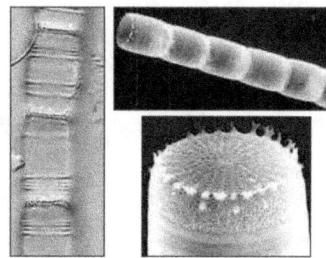

Photographs 6–8: *Stephanodiscus binderanus*, a fiamentous centric diatom. Photographs courtesy of Rex Lowe, Bowling Green State University.

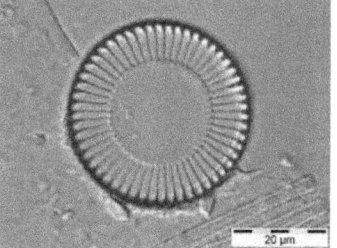

Photograph 9: The centric diatom *Cyclotella meneghiniana*. Photograph courtesy of the Philadelphia Academy of Natural Sciences.

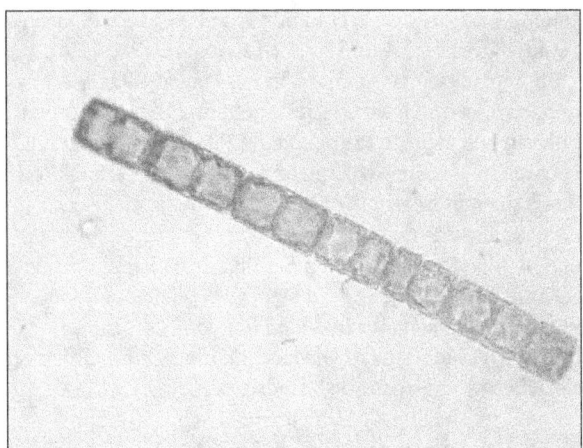

Photograph 10: *Melosira* sp., a centric filamentous diatom. Photograph by Kurt Carpenter, June 26, 2006.

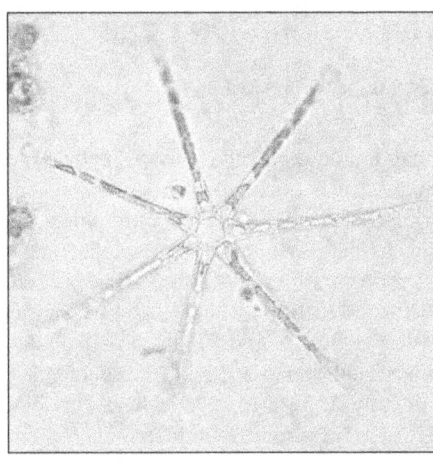

Photograph 11: *Asterionella formosa*, a colonial planktonic diatom. Photograph by Kurt Carpenter, July 16, 2008

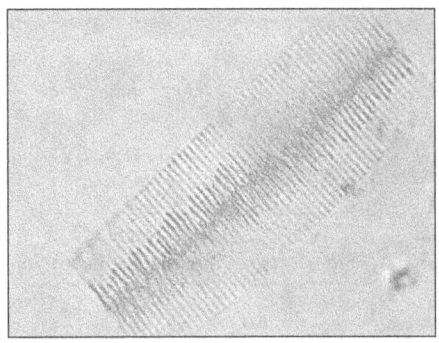

Photograph 12: *Fragilaria crotonensis*, a colonial planktonic diatom. Photograph by Kurt Carpenter, July 7, 2006.

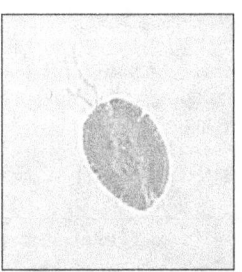

Photograph 13: *Cryptomonas erosa*, a Cryptophyte unicellular flagellate. Photograph by Kurt Carpenter, June 30, 2008.

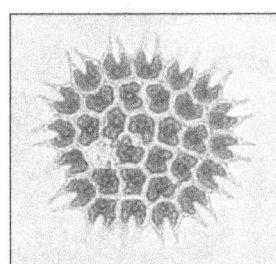

Photograph 14: *Pediastrum*, a colonial Green algae. Photograph by Kurt Carpenter, July 24, 2006.

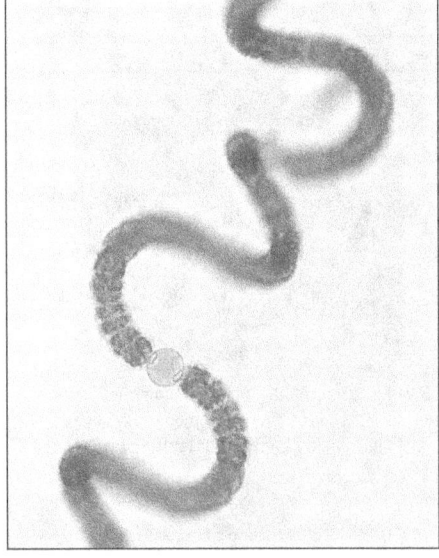

Photograph 15: *Anabaena flos-aquae*, a colonial blue-green algae. Photograph by Kurt Carpenter, July 15, 2008.

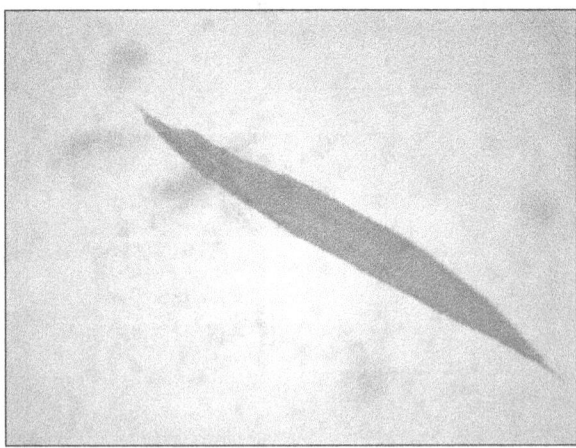

Photograph 16: *Aphanizomenon flos-aquae*, a colonial blue-green algae. Photograph by Kurt Carpenter, July 16, 2008.

Diatoms and green algae comprised 85 percent of the total number of algal taxa (93 and 28 taxa, respectively), and blue-green and golden algae contributed 6 taxa each. While most of the algae identified in Tualatin River were planktonic and characteristic of lake phytoplankton (Porter, 2008), some were facultatively planktonic, meaning that they can reside on the bottom and migrate to the water column when conditions permit (Wehr and Sheath, 2003). Colony-forming algae were common, including the blue-green algae *Anabaena* and *Aphanizomenon*, green algae *Scenedesmus quadricauda, Pediastrum*, and *Actinastrum hantzschii*, and the diatoms *Asterionella formosa* and *Fragilaria crotonensis*. A few benthic taxa also were a part of the phytoplankton assemblage, tending to occur most often at the upstream sites (Rood Bridge and RM 24.5 near Scholls), at relatively low abundances, especially once flows declined.

Most of the algal taxa, and the majority of the most abundant taxa, are considered "eutrophic" and either require or prefer elevated nutrient concentrations (table 6), which is consistent with the relatively high concentrations of nitrate-nitrogen and moderate SRP concentrations in the river most of the time. Although several taxa were consistently abundant in samples collected during all 3 years, about one third, or 53 of the 143 algal taxa, occurred in just one of three years, highlighting the annual variability in species composition. Most (29) of these unique taxa were identified in 2008, whereas 10–15 taxa were unique to 2006 and 2007. Just 7 of the 29 taxa unique to the 2008 growing season were identified in drainage water from Wapato Lake (see section, "Wapato Lake Algal Bloom"). These data indicate that although there is some consistency in phytoplankton species composition from year to year, particularly among the most abundant species, considerable variations in abundance and character occur.

The most abundant algal taxa in 2006–08 was *Cryptomonas erosa*, a relatively small (520 μm^3) flagellated alga which comprised 33 percent of the overall total algal biovolume for all sites over the 3-year period (see photograph 13, p. 28). This taxa occurred in 97 percent of phytoplankton samples (table 6). Only the red algae *Rhodomonas minuta*, another relatively small flagellate, was more frequently detected (99 percent of samples). Other important taxa included two *Stephanodiscus* taxa (*S. binderanus* [see photographs 6–8, p. 27], and *S. hantzschii*), which comprised 21 percent of the overall total biovolume, the blue-green *Anabaena flos-aquae,* which contributed 11 percent (nearly all in 2008), and the green algal

flagellate *Chlamydomonas* sp. contributed 7 percent. These 5 taxa accounted for 71 percent of the total phytoplankton biovolume among all samples.

Blue-green algae, either *Aphanizomenon, Anabaena,* or both (see photographs 15–16, p. 28) were identified in the Tualatin River during USGS longitudinal samplings each summer in 2006–08, but their abundance was highly variable (appendix B). *Microcystis*, another type of blue-green algae that often was reported to be present in the Tualatin River decades earlier (Oregon Department of Environmental Quality and Unified Sewerage Agency, 1982), was only identified in July 2008 during the large *Anabaena* bloom, and only in fresh samples observed microscopically; none were identified in preserved samples. These blue-green algae species tend to cloud the water or form surface scums during quiescent conditions, and many strains produce potent liver or neural toxins that can be a threat to public health.

A comparison of the phytoplankton species identified in the Tualatin River during 2006–08 with previous USGS samples collected in 1991–93 showed that the overall assemblages were quite similar, with 78 percent of the taxa detected previously. In addition, nearly half (49 percent) of the algal species identified during the current study also were identified in samples from Hagg Lake between 2000 and 2006 (Rounds and others, 1999; Bonn, 2006). Two taxa, *Stephanodiscus binderanus* and *Cyclotella pseudostelligera*, were absent from the 1991–93 collections from the Tualatin River, whereas they were abundant in 2006–08, especially *S. binderanus*. According to Kipp and others (2013), this species was imported from Eurasia to the Great Lakes in ballast water in 1938 and is considered nonindigenous. *S. binderanus* occurs in eutrophic conditions, tolerates a wide range of water temperature and osmotic pressure, and can be a nuisance by causing taste and odor problems in drinking-water supplies and by clogging filters.

The phytoplankton species composition in the Tualatin River is dynamic and shows a general seasonal succession. Despite year-to-year variation, the filamentous centric diatom *Stephanodiscus binderanus* dominated or co-dominated the early summer bloom each year during 2006–08 along with other diatoms as river flow was decreasing. The initial bloom of *S. binderanus* typically was followed by a bloom of small flagellates (*Cryptomonas* and *Chlamydomonas*), then by a different centric diatom (*Melosira granulata* or *Stephanodiscus hantzschii*) in 2006–07, or in 2008 by a large bloom of the blue-green algae *Anabaena flos-aquae* (appendix B).

Table 6. Dominant algal taxa in the Tualatin River, Oregon, 2006–08.

[Includes top 20 taxa by percent occurrence and top 20 taxa by total biovolume for all samples. **Abbreviations:** RM, river mile; 99W, Highway 99 west; d/s, downstream; ?, unknown autecology]

Scientific name	Species code	Division	Growth habit	Trophic indicator	Samples (percent)	Number of detections 2006	2007	2008	Wapato	Wapato pump discharge[1]	Spring Hill pump station[1]	Food Bridge	RM 24.5	99W/ Jurgens	d/s Cook Park	d/s Fanno Creek	Stafford Road	Oswego Dam
Rhodomonas minuta	RHodminu	Cryptophyte	Planktonic	?	99	34	37	54	2	14,365	2,148	2,199	1,251	25,280	16,364	11,817	32,192	15,089
Cryptomonas erosa	CRYerosa	Cryptophyte	Planktonic	Eutrophic	97	33	37	52	2	4,311,115	418,786	50,371	74,646	285,659	361,073	216,529	476,395	629,238
Stephanodiscus hantzschii	SThant	Diatom	Planktonic	Eutrophic	97	32	37	51	1	72,160	15,463	14,918	11,255	71,264	78,922	60,516	77,700	122,838
Cyclotella pseudostelligera	CYpsdste	Diatom	Planktonic	Eutrophic	93	31	36	47	1	38,110	1,745	1,083	3,642	38,863	45,173	42,533	33,901	44,922
Chlamydomonas sp.	CHLsp	Green	Planktonic	Eutrophic	82	26	33	46	2	570,014	52,348	14,658	14,583	63,779	63,256	63,030	142,110	66,275
Ankistrodesmus falcatus	ANKfalca	Green	Planktonic	Eutrophic	82	23	27	54	2	61,386	4,027	979	4,153	3,679	2,668	2,229	3,379	2,286
Stephanodiscus binderanus	STbinder	Diatom	Planktonic	Eutrophic	75	25	32	34	0	0	0	4,535	5,675	146,384	168,135	139,399	256,172	136,818
Selenastrum minutum	SELminut	Green	Planktonic	Eutrophic	70	16	29	42	0	0	0	505	917	1,135	1,200	1,271	585	483
Unidentified flagellate	XXsp	Unknown	Planktonic	?	59	26	22	27	2	1,774,045	5,369	449	893	528	1,162	3,389	0	0
Scenedesmus quadricauda	SCEquadr	Green	Planktonic	Eutrophic	55	12	15	45	2	385,981	65,784	5,051	10,330	12,611	12,631	12,259	14,417	6,808
Cyclotella meneghiniana	CYmenegh	Diatom	Planktonic	Eutrophic	53	21	16	30	1	57,127	10,201	2,228	6,004	24,490	20,182	21,218	6,384	20,863
Chromulina sp.	CHRsp	Chrysophyte	Planktonic	?	53	16	19	30	0	0	0	0	232	431	510	917	607	961
Nitzschia acicularis	NZacicul	Diatom	Planktonic	Eutrophic	41	13	10	29	1	77,951	7,517	5,591	3,698	3,601	6,445	2,579	4,925	20,244
Gomphonema angustatum	GOangstt	Diatom	Benthic	Oligotrophic	35	6	13	28	0	0	4,832	4,589	1,259	1,127	709	1,045	795	5,188
Melosira distans alpigena	MEdist-alp	Diatom	Planktonic	?	38	11	15	20	0	0	0	592	2,175	4,951	5,322	6,203	10,965	10,613
Sphaerocystis schroeteri	SPHschro	Green	Planktonic	Eutrophic	36	11	9	25	1	77,951	22,550	531	5,887	6,722	14,206	10,587	7,276	5,359
Achnanthes lanceolata	AClanc	Diatom	Benthic	Eutrophic	29	5	12	20	0	0	0	5,464	1,311	1,451	1,800	249	271	1,345
Glenodinium sp.	GLenosp	Dinoflagellate	Planktonic	?	30	13	13	10	0	0	0	3,332	0	19,714	14,017	8,849	17,076	14,702
Crucigenia quadrata	CRUquad	Green	Planktonic	Eutrophic	28	8	4	23	1	12,778	2,282	532	666	1,581	555	1,437	2,628	3,972
Achnanthes minutissima	ACminu	Diatom	Benthic	Eurytrophic	25	7	7	20	0	0	1,342	1,507	428	486	97	301	58	120
Stephanodiscus astraea minutula	Stas-mi	Diatom	Planktonic	?	25	12	8	10	0	0	0	692	307	4,325	1,973	3,569	14,143	10,968
Anabaena flos-aquae	ABFA	Bluegreen	Planktonic	Eutrophic	17	0	0	26	2	287,248	89,932	167,588	32,051	235,648	111,138	94,823	48,178	0
Aphanizomenon flos-aquae	AFNfloaq	Bluegreen	Planktonic	Eutrophic	15	2	3	18	1	87,694	1,691	164,283	3,964	5,391	22,019	29,526	2,107	92,859
Melosira granulata	MEgran	Diatom	Planktonic	Eutrophic	14	7	5	5	0	0	0	0	1,075	1,036	4,204	542	30,741	61,093
Synedra ulna	SYulna	Diatom	Benthic	Eurytrophic	8	3	1	9	0	0	0	30,760	2,492	0	0	2,946	0	14,159
Actinastrum hantzschii	AThn	Green	Planktonic	?	6	0	1	8	1	267,259	0	0	0	10,177	15,825	7,959	16,443	7,159
Melosira ambigua	MEambigu	Diatom	Planktonic	Eutrophic	6	5	1	1	0	0	0	0	0	4,791	16,373	0	36,894	1,460
Fragilaria crotonensis	FRcroton	Diatom	Planktonic	Mesotrophic	4	4	1	0	0	0	0	1,056	78,228	0	0	0	0	8,240

[1] In 2008, two samples were collected from the Wapato pump discharge and one sample from the Spring Hill pump station.

Small flagellates, particularly *Cryptomonas erosa*, were secondary dominant taxa during the early part of the summer, when other taxa such as *S. binderanus* and two eutrophic *Melosira* taxa (*M. ambigua* and *M. italica*) were dominant. Large peaks in biovolume were measured for *Melosira ambigua* (at Stafford Road, July 2006) and *Stephanodiscus binderanus* (several sites, July 2007 and June 2008). These blooms were followed by sharp declines in algal biovolume between late July and early August at Stafford Road (RM 5.4) in 2006 and 2007, mostly from declines in the abundance of diatoms (fig. 16), especially *Stephanodiscus binderanus*, whereas the relative biovolume of *Cryptomonas erosa* increased from 5–25 percent to over 80 percent. During this late July to early August period, some colonial forms of green algae such as the spined *Scenedesmus quadricauda* became proportionally more abundant, although its overall abundance declined along with that of the other types of algae.

Conditions were considerably different in 2008, when the draining of Wapato Lake (near RM 60) provided a rich inoculum of blue-greens and other algae, copepods, and high levels of bioavailable nutrients—particularly phosphorus—that supported a large bloom of *Anabaena flos-aquae* downstream. It is interesting that the decline in diatoms, which was observed during all three summers, occurred 2–3 weeks earlier in 2008 than in 2006–07. The diatom decline may have been precipitated by competition from other types of phytoplankton (blue-green algae) and/or enhanced grazing pressures from copepods and other zooplankton that were discharged from the Wapato Lake agricultural area to the upper river.

Wapato Lake Algal Bloom

During July 2008, a large bloom of *Anabaena flos-aquae* (primarily) and *Aphanizomenon flos-aquae* occurred in the lower Tualatin River, prompting the Oregon Health Authority (formerly the Department of Human Services) to issue a public health advisory for recreational water contact on July 12 for the reach extending from Jurgens Park (RM 10.8) to the river mouth (Bonn, 2008). This bloom formed a thick surface scum along the margins and backwater areas for several miles of the lower Tualatin River. *Anabaena* can produce potent nerve and liver toxins (anatoxin-*a* and microcystin), and although tests were not conducted for the more toxic anatoxin-*a*, microcystin was detected at concentrations ranging from 0.14 to 2.4 µg/L. Although low levels of these potentially toxic blue-green algae were observed in the river each year, cell counts in 2006 and 2007 were much lower than those in 2008.

The July 2008 bloom was quickly traced upstream to the Wapato Lake agricultural area located adjacent to and southeast of Gaston, Oregon (fig. 2). Because of a breach in one of the levees during a high-flow event on December 2–3, 2007, the low-lying area was flooded to a much greater extent than normal and could not be drained until much later than normal in spring and summer 2008. Drainage water entering the Tualatin River near RM 60 was elevated in ammonia and SRP and contained a rich inoculum of algae including *Anabaena flos-aquae* and *Aphanizomenon flos-aquae*, several types of flagellates, including *Trachelomonas volvocina*, *Cryptomonas erosa*, *Chlamydomonas* sp., green algae including *Scenedesmus*, *Actinastrum*, and *Sphaerocystis*, and many types of diatoms, including planktonic *Cyclotella meneghiniana* and *C. pseudostelligera*. Drainage water from Wapato Lake also contained very high abundances of zooplankton (appendix C), primarily copepods, which continued to thrive in the lower Tualatin River during the bloom, and possibly affecting the algal population through selective grazing.

Zooplankton

The diverse phytoplankton assemblage in the lower river sometimes supports sizable populations of zooplankton—including protozoans, rotifers, copepods, and cladocerans. These organisms feed on a wide variety of foods such as algae, bacteria, detritus, and other zooplankton, by selective and indiscriminant filter feeding (cladocerans and rotifers), and by direct grazing (copepods) (Wetzel, 1983).

The most abundant zooplankton identified in Tualatin River samples included early life stages of copepods (nauplii and cyclopoid copepodites), the cladocerans *Bosmina longirostris* and *Chydorus sphaericus*, and the copepod *Diacyclops thomasi* (table 7) (see photographs 17–19, p. 34). These taxa were identified in 70–100 percent of samples. The most abundant zooplankton taxa, by far, was *Bosmina longirostris*, which made up nearly 28 percent of the total density considering all samples collected during the study and was particularly abundant in the lower reaches of the river. This taxon reached its highest density in late June 2007, when the density at Stafford Road exceeded 60,000 individuals per cubic meter (appendix C).

Total zooplankton densities in the Tualatin River were much higher in 2008 than in 2006–07 for sites extending from Rood Bridge (RM 38.4) to Highway 99W/Jurgens Park (RM 11.6–10.8). Water discharged from the Wapato Lake agricultural area in 2008 contained high abundances of zooplankton and was an important source of zooplankton to the river that year.

Zooplankton populations in the Tualatin River exhibited important longitudinal and seasonal patterns during 2006–08. Zooplankton densities showed a distinct longitudinal increase, particularly downstream of Highway 99W (RM 11.6), about 20 mi downstream of where the reservoir reach begins (fig. 17). A seasonal increase in zooplankton abundance was observed in early summer (fig. 18) just as or shortly after the phytoplankton population began to increase in the river (fig. 19). These patterns, especially for the lower river sites, are a typical result consistent with phytoplankton being a major food source for zooplankton.

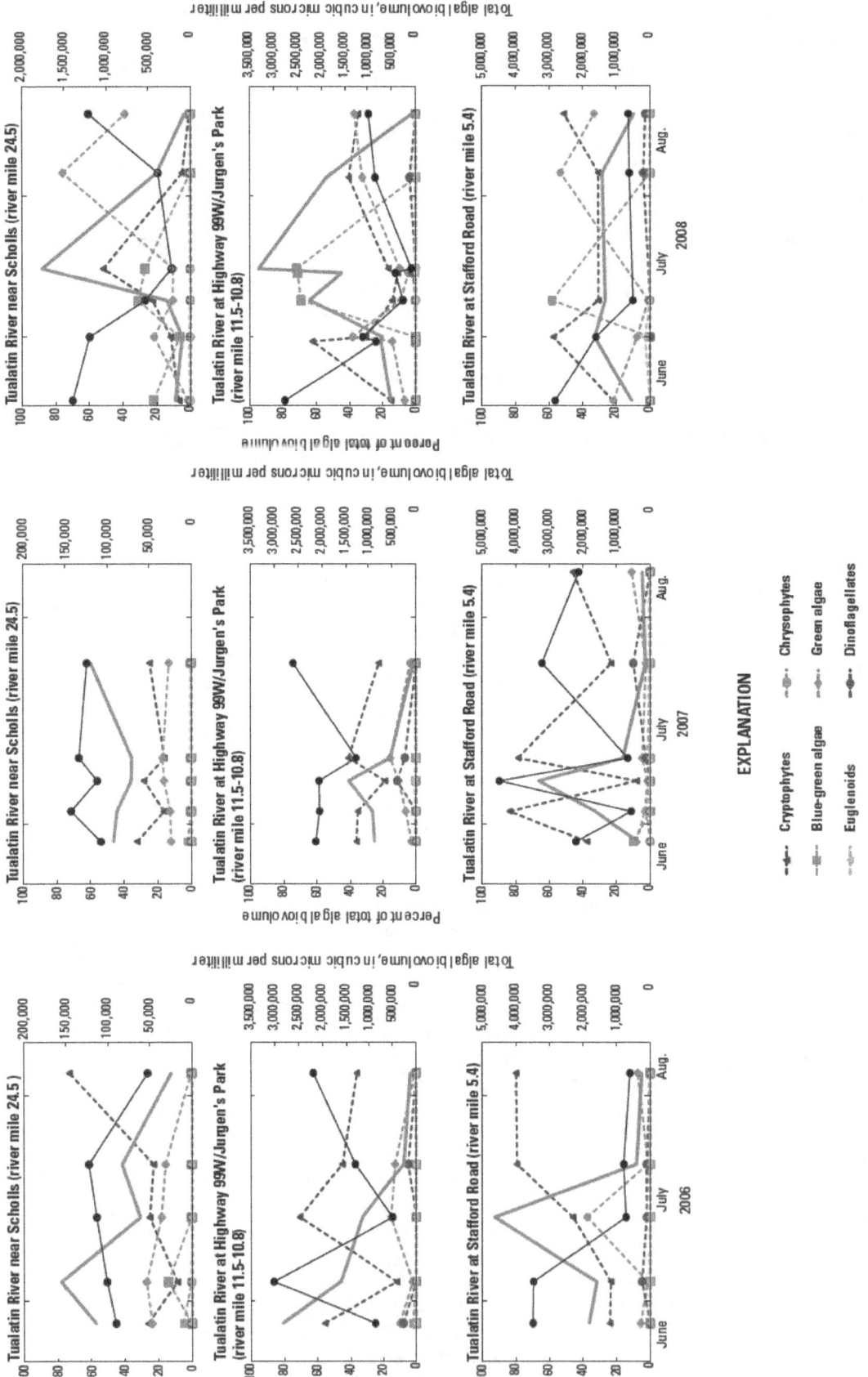

Figure 16.   Time series of total algal biovolume and percentage of major phytoplankton groups (divisions) at selected sites in the Tualatin River, Oregon, 2006–08.

Table 7. Dominant zooplankton taxa in the Tualatin River, Oregon, 2006–08.

[Includes top 20 taxa by percent occurrence and top 20 taxa by total density for all samples. **Abbreviations:** RM, river mile; 99W, Highway 99 West; d/s, downstream]

| Zooplankton taxa | Group | Samples (percent) | Number of occurrences | | | | Average density (number per cubic meter) | | | | | | | | |
			2006	2007	2008	Wapato	Wapato pump discharge[1]	Spring Hill pump station[1]	Rood Bridge	RM 24.5	99W Jurgens Park	d/s Cook Park	d/s Fanno Creek	Stafford Road	Oswego Dam
Copepod nauplii	Copepoda	100	24	33	44	2	36,076	0	293	554	586	1,897	667	933	1,853
Cyclopoid copepodites	Copepoda	97	22	34	42	2	400,148	6,780	87	160	2,259	2,824	1,067	512	687
Bosmina longirostris	Cladocera	94	23	32	40	2	3,867	0	18	49	414	2,693	1,771	6,290	4,937
Chydorus sphaericus	Cladocera	80	18	28	35	1	704	0	105	4	37	83	58	113	55
Chironomidae sp.	Other zooplankters	79	24	24	32		0	0	10	10	15	26	35	21	25
Diacyclops thomasi	Copepoda	70	12	24	35	2	277,737	6,780	34	22	130	431	331	95	75
Euchlanis dilatata	Rotifera	66	15	20	32		0	0	15	13	30	9	20	31	21
Ceriodaphnia dubia	Cladocera	65	14	23	29		0	0	6	5	22	256	105	234	1,078
Synchaeta sp.	Rotifera	60	17	18	26		0	0	5	1	2,844	75	39	509	124
Brachionus angularis	Rotifera	53	4	19	31	1	877	0	15	45	64	272	84	43	30
Scapholeberis armata	Cladocera	50	14	14	23		0	0	1	2	70	170	39	78	225
Philodina sp.	Rotifera	47	6	19	22		0	0	3	2	2	7	10	14	6
Harpacticoid copepods	Copepoda	46	7	17	22		0	0	5	2	6	14	12	195	4
Diaptomus reighardi	Copepoda	45	5	20	20	1	4,930	0	4	2	21	32	26	212	55
Diaptomus sp. *copepoidites*	Copepoda	45	7	21	17		0	0	3	2	24	31	9	64	39
Microcyclops varicans	Copepoda	45	11	18	16		0	0	12	4	2	15	9	4	36
Difflugia sp.	Other zooplankters	42	8	14	20	2	9,489	1,695	5	6	81	31	12	9	22
Polyarthra vulgaris	Rotifera	42	11	10	21		0	0	78	33	11	31	1	20	18
Brachionus havanensis	Rotifera	38	12	19	7		0	0	5	4	54	1,573	295	60	37
Brachionus quadridentata	Rotifera	30	0	15	15		0	0	0	253	1,322	590	179	92	21

[1]In 2008, two samples were collected from the Wapato pump discharge and one sample from the Spring Hill pump station.

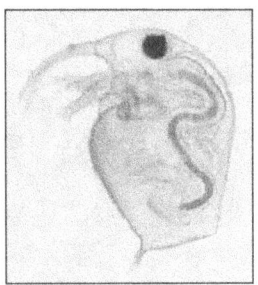

Photograph 17: *Bosmina longirostris,* a cladoceran. (Photograph by Kurt Carpenter, June 26, 2006.)

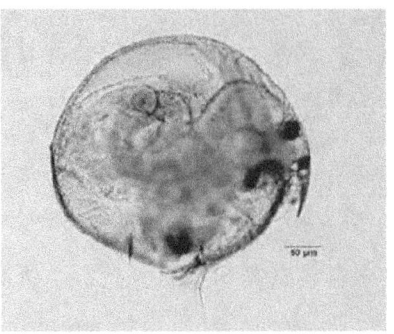

Photograph 18: *Chydorus sphaericus,* a cladoceran. (Photograph by WIM VAN EGMOND/VISUALS UNLIMITED, INC. /SCIENCE PHOTO LIBRARY.)

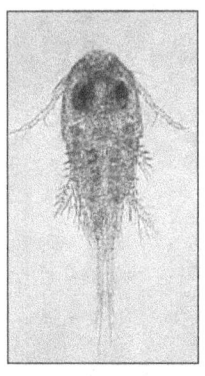

Photograph 19: The copepod *Diacyclops thomasi* (Photograph by Kurt Carpenter, August 6, 2008.)

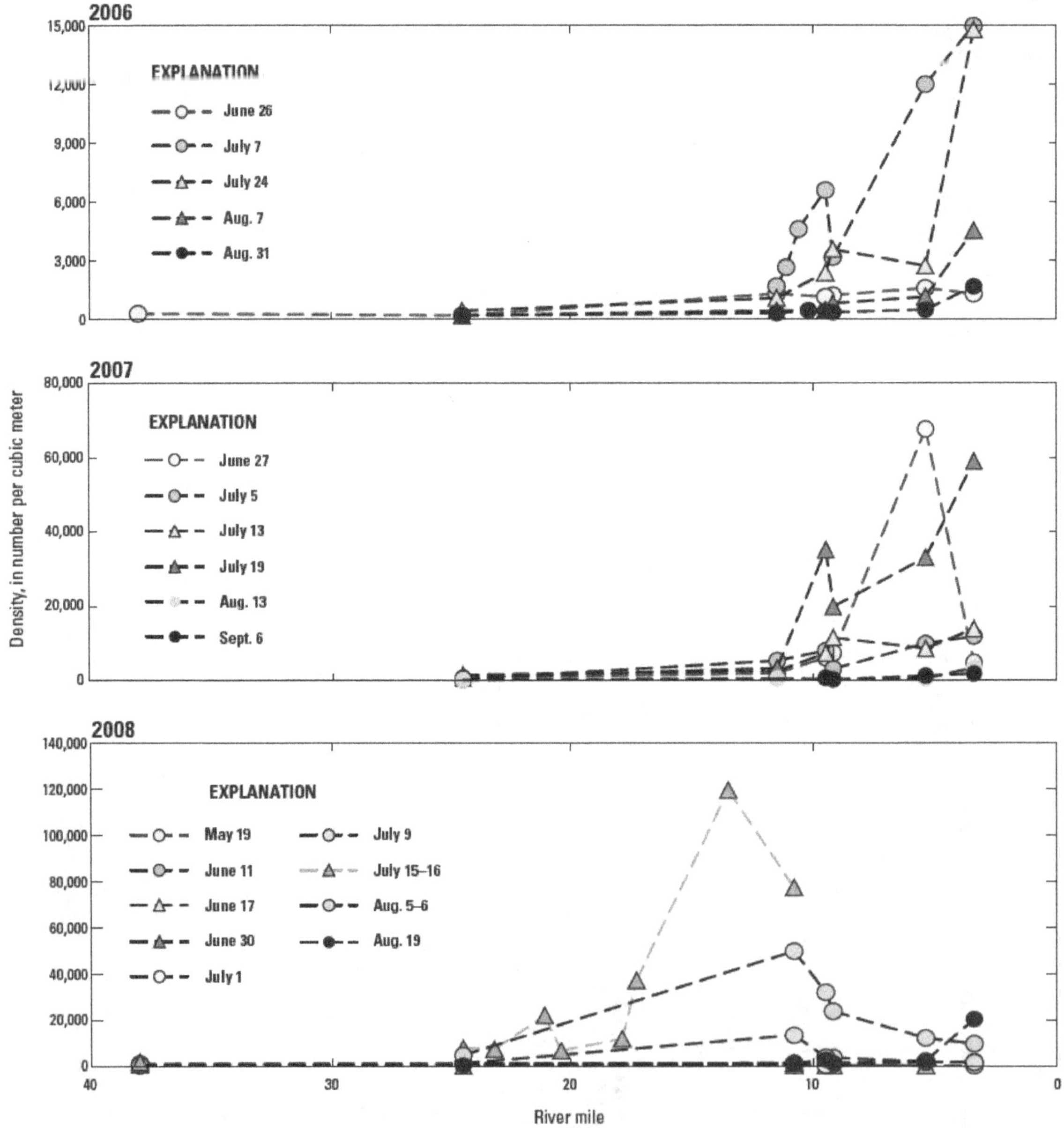

Figure 17.   Longitudinal pattern in total zooplankton density in the Tualatin River, Oregon, 2006–08. Note the variable y-axes scales.

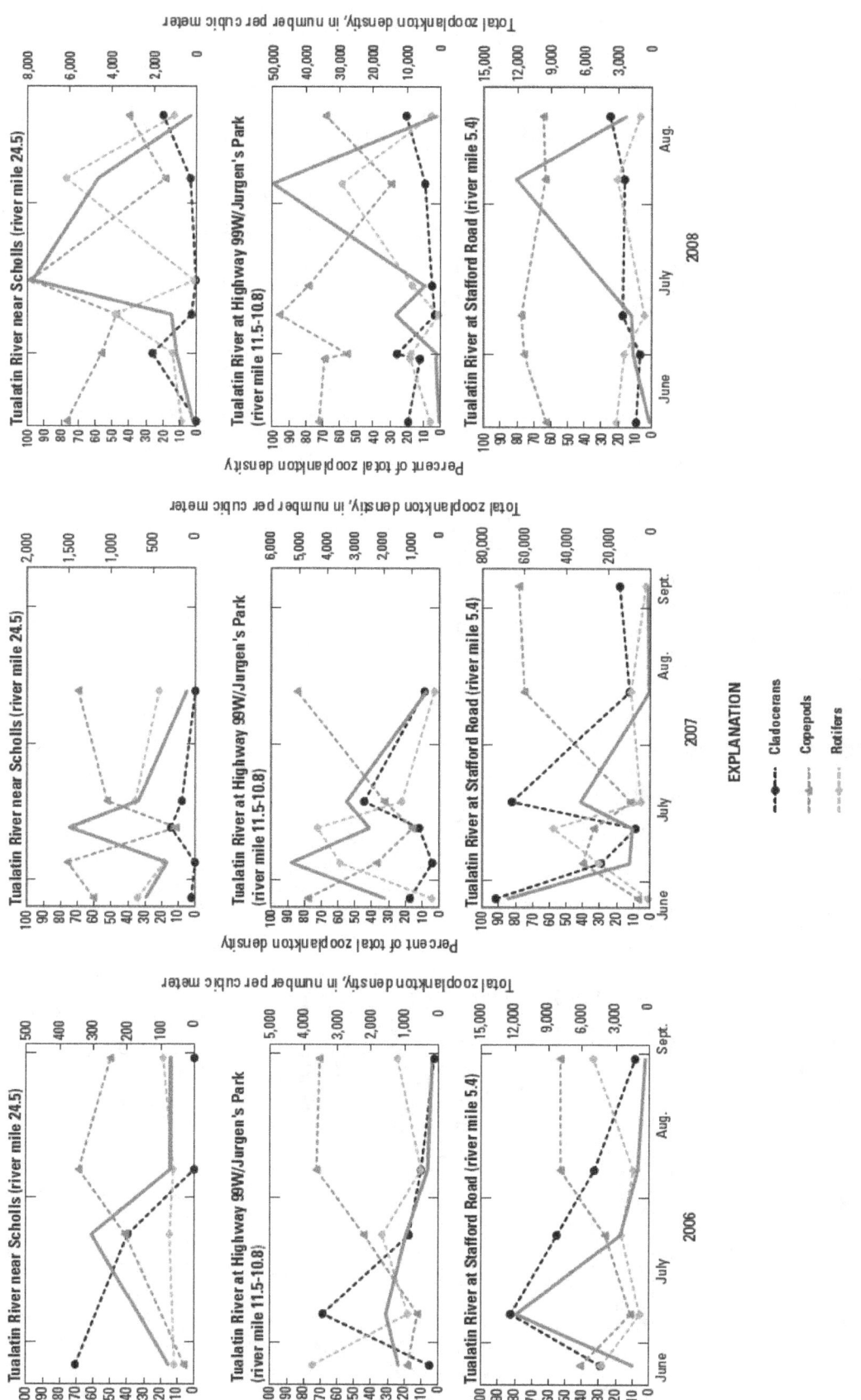

Figure 18. Seasonal pattern in zooplankton abundance and percentage of major groups at select sites in the Tualatin River, Oregon, 2006–08. Note the variable y-axes scales.

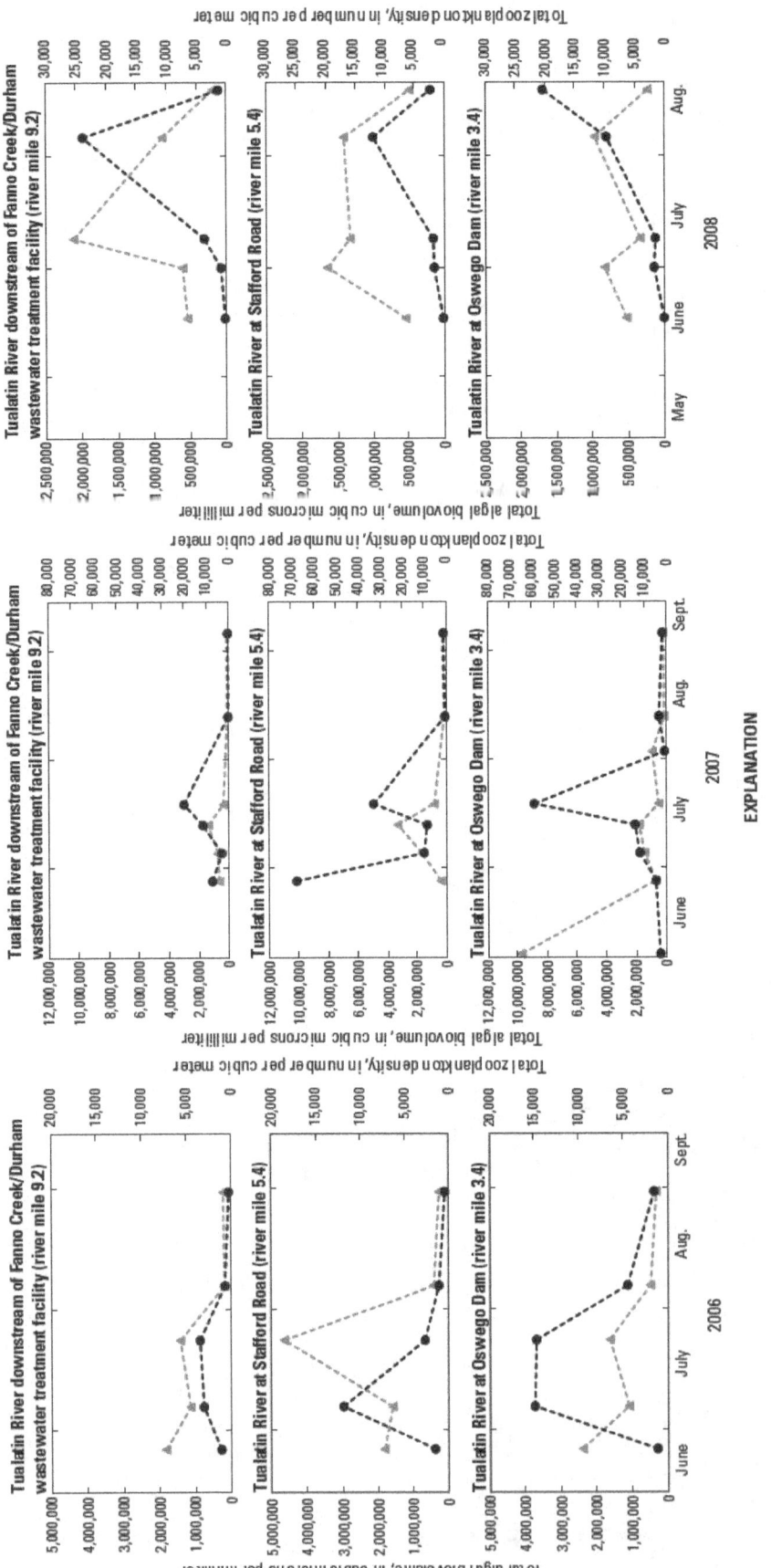

Figure 19. Time series of algal biovolume and zooplankton density at select sites in the Tualatin River, Oregon, 2005–08. Note the variable y-axes scales.

The seasonal successional pattern in zooplankton indicates that cladocerans tended to follow the major peak in the total zooplankton density, decreasing in abundance after the decline in algal populations. After that point, when the overall density of zooplankton decreased, copepods became the dominant zooplankton in the river.

Multivariate Analyses of Phytoplankton Assemblages and Environmental Data

A combination of multivariate data-analysis techniques were performed to discern patterns in the phytoplankton species data, to test the relative importance of environmental factors in explaining variations or patterns in the phytoplankton species composition, and to identify which species were most important in explaining differences between selected groups of samples.

Nonmetric Multidimensional Scaling (NMDS) Ordinations

The 2006–08 plankton dataset consisted of 143 phytoplankton and 99 zooplankton taxa. Considering only those taxa occurring in more than 3 samples resulted in a total of 78 phytoplankton taxa. NMDS ordinations of the phytoplankton assemblage data were used to graphically represent the patterns among samples; this identified four subsets of samples for further analysis (described below). By first determining the patterns in the data and how samples may be related, a better understanding of the factors that influence the composition, abundance, and dynamics of phytoplankton communities in the Tualatin River can be obtained.

The ordination analysis first suggested a longitudinal separation of sites, with samples from the two uppermost Tualatin River sites (Rood Bridge at RM 38.4 and RM 24.5 near Scholls) separating from the downstream sites in the reservoir reach (fig. 20A). Algal abundance (Chl-*a* and total biovolume) at these sites tended to be lower than sites farther downstream, with relatively greater abundances of benthic diatoms including *Gomphonema angustatum*, *Achnanthes lanceolata*, and *Synedra ulna* (table 6). Water temperatures also were almost always lower at these upstream sites compared with the downstream sites.

Phytoplankton samples collected in late June and early July 2008 during the *Anabaena* bloom associated with the draining of Wapato Lake formed another clustered group of samples in the NMDS ordination (fig. 20B). This separation was determined largely by a greater abundance of blue-green algae (*Anabaena* sp. and *Aphanizomenon flos-aquae*), green algae (*Scenedesmus quadricauda*, *Scenedesmus acuminatus*,

Chlamydomonas sp., and *Sphaerocystis schroeteri*), and the high-phosphorus-indicating diatom *Cyclotella meneghiniana* in the Wapato-affected samples. Algal taxa with higher abundances in the non-Wapato-affected sample group included two eutrophic centric diatoms, *Stephanodiscus binderanus* and *Cyclotella pseudostelligera*. Algal taxa having relatively high abundance in both groups included *Cryptomonas erosa*, *Rhodomonas minuta*, and *Stephanodiscus hantzschii*. These three taxa were nearly ubiquitous in the Tualatin River, occurring in 97–100 percent of samples at varying abundances.

After removing the more upstream and Wapato-affected samples, the remaining lowermost-basin samples separated along the first ordination axis (fig. 20C), which reflects the seasonal change in the phytoplankton species composition from May–July to August–September. Filamentous centric diatoms *Stephanodiscus binderanus* and *S. hantzschii* were more abundant in May–July, whereas small flagellates (*Cryptomonas erosa* and *Chlamydomonas* sp.) and a number of green algal taxa were relatively more abundant in August and September (table 8).

Separation of these late summer samples into distinct groups based on algal community data coincided with several changes in river conditions from July to August. Phytoplankton populations tended to decline sharply during this time interval, and zooplankton densities, which were relatively high in July, plummeted in early August (fig. 19), resulting in a "clear water phase" for a period lasting 3–5 weeks. This also is the time period when natural flows reached their minimum for the season, and flow augmentation and WWTF effluent were at their highest percentage (fig. 12).

Spearman Rank Correlations, PCA, and BEST Analyses

Spearman rank correlations and Principal Components Analyses (PCA) were used to identify sources of variation in the datasets to better understand the underlying data structure. The Spearman rank correlation matrix for the full suite of environmental variables listed in table 9 and for groups of algae and zooplankton (for non-Wapato-affected lower Tualatin River samples over the 3-year study period) is given in table 10. Although many of the correlations in these tables are significant, care should be exercised before drawing firm conclusions from these tables because such correlations do not necessarily indicate a cause-and-effect relation between variables. Nevertheless, many of the variables, including biovolumes of individual algal divisions and various components of flow, had significant correlations with important water-quality variables, including total flow augmentation (total reservoir releases minus withdrawals) and DO concentrations (table 10).

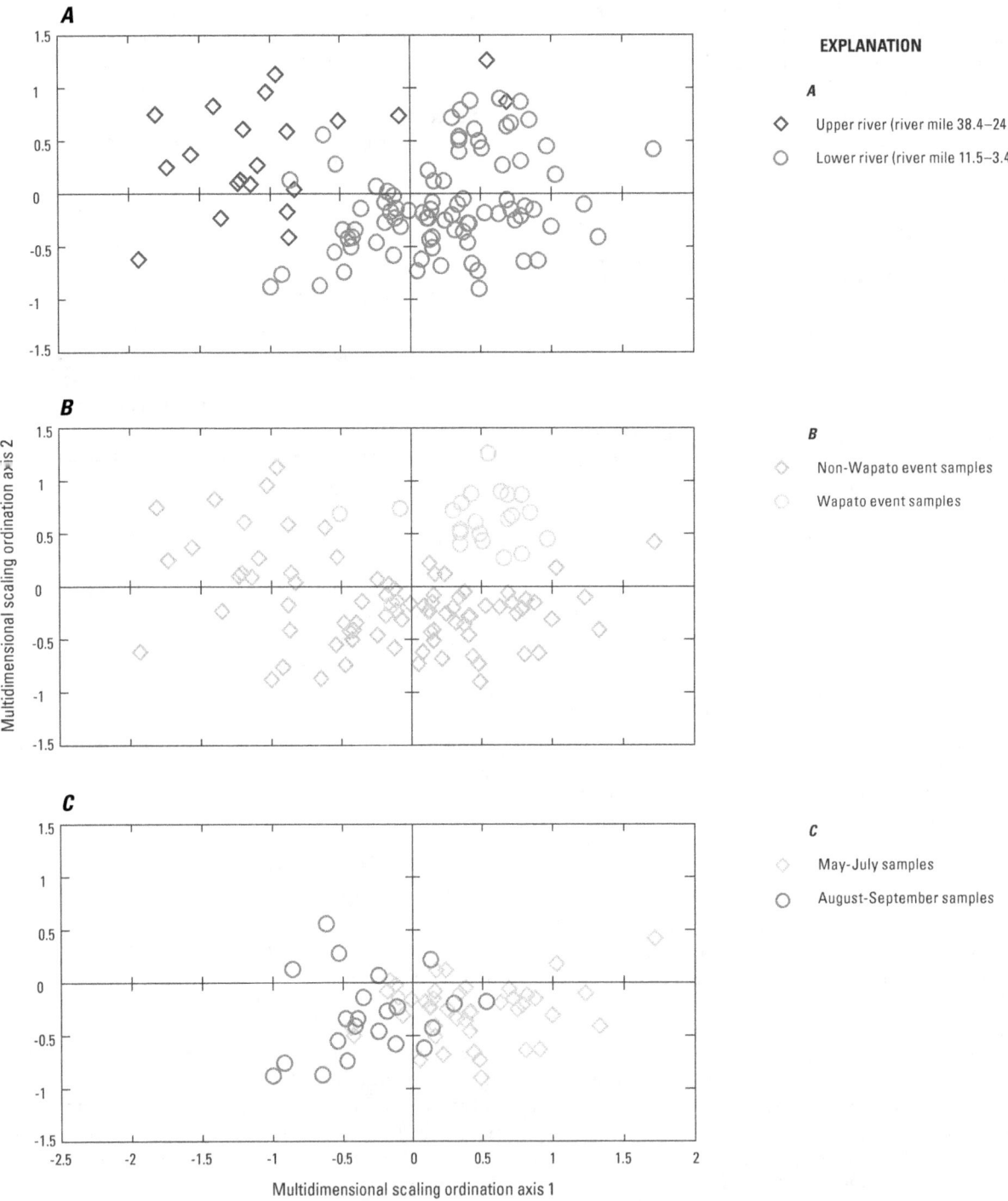

EXPLANATION

A

◇ Upper river (river mile 38.4–24.5)

○ Lower river (river mile 11.5–3.4)

B

◇ Non-Wapato event samples

○ Wapato event samples

C

◇ May-July samples

○ August-September samples

Figure 20.   Phytoplankton samples from the Tualatin River, Oregon, 2006–08, highlighting (*A*) the two upstream sites, (*B*) the group of samples collected during and after the *Anabaena flos-aquae* bloom associated with the draining of Wapato Lake, and (*C*) downstream sites with Wapato-affected samples removed, showing separation of samples collected in May–July versus August–September.

Table 8.   Average phytoplankton percent biovolume for non-Wapato affected lower river samples for May–July and August–September 2006–08.

[Results from Primer-E SIMPER analysis showing species that were important for separating the May–July samples from August–September samples]

Algal taxa	Algal division	May–July (percent)	August–September (percent)
Higher in May–July			
Stephanodiscus binderanus	Diatom	25	7
Stephanodiscus hantzschii	Diatom	10	3
Higher in August–September			
Cryptomonas erosa	Cryptophyte	35	44
Less than 5 percent change			
Cyclotella pseudostelligera	Diatom	6	5
Aphanizomenon flos-aquae	Bluegreen	1	0
Melosira ambigua	Diatom	1	0
Nitzschia acicularis	Diatom	1	0
Chlamydomonas sp.	Green	6	8
Cyclotella meneghiniana	Diatom	1	4
Actinastrum hantzschii	Green	0	3
Sphaerocystis schroeteri	Green	0	2
Ankistrodesmus falcatus	Green	0	1
Crucigenia quadrata	Green	0	1
Melosira distans alpigena	Diatom	1	2
Melosira granulata	Diatom	1	2
Scenedesmus quadricauda	Green	1	2
Synedra ulna	Diatom	0	1
Tetrastrum staurogeniaforme	Green	0	1
Rhodomonas minuta	Cryptophyte	2	3

BEST analyses were performed to determine which combination of environmental variables listed in table 9 could explain the greatest amount of variation in the phytoplankton assemblage structure. Nine significant (P<0.001) BEST solutions were produced: three for the entire 3-year study period and six for the individual years (table 11).

Often, the environmental variables that were most highly correlated to algal assemblages were variables that were dependent on algal conditions (biomass and metabolism), including Chl-*a*, algal biovolume, DO, DO percent saturation, and pH. Because the objective of the BEST analysis is to understand the independent factors that shape algal assemblages, these algal-dependent biomass and metabolism parameters were removed from the pool of variables for the analysis. A similar dependency exists between algae and dissolved nutrients, particularly ammonia and SRP. Although previous studies have identified the importance of phosphorus in regulating phytoplankton populations in the Tualatin River, and enriched nutrient concentrations help to explain the prevalence of eutrophic taxa in the river (table 6), uptake of nutrients by algae is great enough that the lowest nutrient

concentrations during summer occur during the largest algal blooms. Therefore, in the cases where SRP was included in a BEST solution, additional BEST analyses were performed without SRP to determine whether other variables emerged as also being significant (table 11).

The individual variables selected for each BEST analysis were determined by the rho value for each variable, and combinations of these variables were selected to maximize the overall rho score. Taken together, the BEST solutions indicate that many of the flow variables—total flow and individual source of flow, including natural flow, flow augmentation, and WWTF effluent—had the most significant influence over algal assemblages in the Tualatin River. Other variables that were sometimes important included chloride or conductance, rotifer and copepod abundance, nitrite-plus-nitrate, and water temperature; and several of those variables were highly correlated with some of the flow variables such as percentage WWTF effluent.

Separating the datasets by year, BEST analyses identified many of the same variables as being important in shaping phytoplankton assemblages as those included in the BEST solutions for the all-year analysis. In 2006, total flow, nitrite-plus-nitrate, rotifer abundance, percent WWTF effluent (or chloride), and water temperature produced the best solution, whereas in 2007, natural flow and flow augmentation, both as a percentage of total flow, were identified as most important, along with SRP, SC, and rotifer abundance. In 2008, total and percentage of natural flow were most important, along with SRP and copepod abundance (table 11).

The similarity percentages (SIMPER) procedure in PRIMER (Clarke and Gorley, 2006) was used to identify the dominant algae during four different algal conditions, including early season periods (Chl-*a* < 10 μg/L); growth phase (Chl-*a* 10–20 μg/L); bloom periods (Chl-*a* > 20 μg/L); and late season (Chl-*a* < 10 μg/L) (table 12). *Stephanodiscus binderanus* and *S. hantzschii* were dominant during the early season, whereas *Cryptomonas erosa*, *S. binderanus*, and *Chlamydomonas* sp. were dominant during the algal "growth phase", when populations were increasing. Two of these taxa, *Cryptomonas erosa* and *S. binderanus*, and *Anabaena flos-aquae* were dominant during bloom periods, and *Cryptomonas erosa* was dominant during the late season, when Chl-*a* concentrations were less than 10 μg/L.

The pairwise analysis of similarity (ANOSIM) tests showed that the largest difference (those with the highest rho values) between one sampling period to the next occurred between the last sample collected in July and the first sample collected each August (table 13), affirming the previous observation of altered algal communities at that time. Recall that the SIMPER analysis of May–July versus August–September samples identified several phytoplankton taxa that were more abundant in each time period (table 8), and the NMDS ordinations also showed a clear distinction in the grouping of phytoplankton samples between these two time periods (fig. 20C).

Table 9.   Summary statistics for environmental variables included in the multivariate analysis of phytoplankton assemblages in the Tualatin River, Oregon.

[Data for 105 main-stem Tualatin River samples collected during 2006–08. **Abbreviations:** WWTF, wastewater treatment facility; µg/L, micrograms per liter; mg/L, milligrams per liter; NTU, nephelometric turbidity units; µS/cm, microsiemens per centimeter; C, degrees Celsius; %, percent; cfs, cubic feet per second [ft^3/s]; PAR, photosynthetically active radiation; (µE/m^2)/s, microeinsteins per square meter per second]

Variable	Definition	Minimum	Median	Maximum	Units
Chl-*a*	Chlorophyll-*a*	1.2	9.2	134	µg/L
Pheo-*a*	Pheophytin-*a*	0.1	2.7	25	µg/L
Pheo-*a*: Chl-*a*	Pheophytin-*a* : chlorophyll-*a* ratio	0.0	0.4	3.5	ratio
DIN	Dissolved inorganic nitrogen (nitrite+nitrate+ammonia)	0.4	3.1	4.8	mg/L
NH$_3$	Dissolved ammonia	0.01	0.027	0.567	mg/L
NO$_2$+NO$_3$	Nitrite+nitrate	0.3	3.1	4.8	mg/L
TKN	Total Kjeldahl nitrogen	0.2	0.6	2	mg/L
TN	Total nitrogen (TKN+nitrite+nitrate)	0.6	3.8	6	mg/L
SRP	Soluble reactive phosphorus	0.006	0.037	0.082	mg/L
TP	Total phosphorus	0.1	0.1	0	mg/L
DIN:SRP	Ratio of DIN to SRP	5.7	83	574	ratio
SRP:TP	Ratio of SRP to TP	0.1	0.4	0.7	ratio
TSS	Total suspended sediment	2.4	6.4	14	mg/L
Turbidity	Turbidity	0.3	6.0	14	NTU
Temp	Water temperature	12.2	21.0	25	°C
SC	Specific conductance	92	254	335	µS/cm
DO	Dissolved oxygen	4.3	7.9	13.7	mg/L
DO%	Dissolved oxygen percent saturation	49	84	163	%
pH	pH	6.3	7.1	8.9	standard units
Cl	Chloride	3.7	17.1	24	mg/L
Total_Flow_cfs	Total streamflow	0	221	587	cfs
Natural_Flow_cfs	Total flow minus reservoir and WWTF releases	0	98	533	cfs
Flow_augmentation_cfs	Total flow from Barney Reservoir and Hagg Lake minus withdrawals[1]	0	69	128	cfs
WWTF_cfs	Total flow from WWTF	0	48	78	cfs
%Flow_augmentation	Percent of flow from Barney Reservoir and Hagg Lake minus withdrawals[1]	0	34	66	percent
%Natural_Flow	Natural flow as percentage of total flow	0	43	100	percent
%WWTF	WWTF discharges as percentage of total flow	0	24	38	percent
PARmax	Maximum daily solar radiation at the Durham WWTF	1,170	1,680	1,850	(µE/m^2)/s

[1] Includes withdrawals from the Hillsboro-Cherry Grove, Patton, Wapato, Tualatin Valley Irrigation District Spring Hill, and Joint Water Commission Spring Hill intakes.

Table 10. Spearman rank correlation matrix for environmental variables and select groups of algae and zooplankton for the lower Tualatin River with Wapato-affected samples removed.

[See table 2 for variable definitions. Rho values in red are significant at the P<0.001 level; Rho values in blue are significant at the P<0.01 level; Rho values in **bold** are significant at the P<0.05 level. Includes 65 samples collected during the 2006–08 growing season from river mile 11.5 to 3.4. **Abbreviation:** <, less than]

Column headers (left to right): Total algal biovolume, Chl-a, Phaeo-a, Phaeo-a:Chl-a, DIN, NH₃, NO₂⁺NO₃, TKN, TN, SRP, TP, DIN:SRP, SRP:TP, TSS, Turbidity, Temp, SC, DO, DO%, pH, Cl, Total_Flow_cfs, Natural_Flow_cfs, Flow_augmentation_cfs, WWTF_cfs, %Flow_augmentation, %Natural_Flow, %WWTF, PARmax, Total zooplankton density, Cladocerans, Copepods, Rotifers, Blue-green algae, Chrysophyte algae, Cryptophyte algae, Diatoms, Centric diatoms, Filamentous diatoms, Dinoflagellates, Euglenoids, Green algae

Row variables (top to bottom): Chl-a, Phaeo-a, Phaeo-a:Chl-a, DIN, NH₃, NO₂⁺NO₃, TKN, TN, SRP, TP, DIN:SRP, SRP:TP, TSS, Turbidity, Temp, SC, DO, DO%, pH, Cl, Total_Flow_cfs, Natural_Flow_cfs, Flow_augmentation_cfs, WWTF_cfs, %Flow_augmentation, %Natural_Flow, %WWTF, PARmax, Total zooplankton density, Cladocerans, Copepods, Rotifers, Blue-green algae, Chrysophyte algae, Cryptophyte algae, Diatoms, Centric diatoms, Filamentous diatoms, Dinoflagellates, Euglenoids, Green algae, Unidentified algae

An oversized version (11 × 17) of this table is available for download at http://pubs.usgs.gov/sir/2013/5037.

Table 11. Summary of BEST analyses listing the top environmental variables explaining patterns in the phytoplankton species composition in the Tualatin River, Oregon, 2006–08.

[See table 9 for varible definitions. **Abbreviations:** BEST, Bio-Env STepwise multivariate analysis; SRP, soluble reactive phosphorus; %, percent; cfs, cubic feet per second (ft^3/s)]

Time period		Sites	BEST variables	Individual variable correlation	BEST solution sample statistics		
					Overall Rho	Significance level	Number of samples
Multi-year							
2006–08 with SRP	May–September	Lower Non-Wapato	SRP	0.242	0.368	P<0.046	65
			%Flow_augmentation	0.215			
			Cl	0.169			
2006–08 without SRP	May–September	Lower Non-Wapato	%Flow_augmentation	0.215	0.27	P<0.001	65
			Natural_Flow_cfs	0.214			
			Rotifers	0.172			
2006–08 without SRP	May–September	Lower Non-Wapato	%Natural_Flow	0.248	0.248	P<0.001	65
2006–08	May–September	All non-Wapato	%WWTF	0.376	0.494	P<0.001	71
			Temp	0.318			
			pH	0.191			
			TSS	0.145			
			Cladocerans and Copepods	0.105			
Individual years							
2006	May–August	Lower	Total_Flow_cfs	0.423	0.574	P<0.001	20
			NO2+NO3	0.371			
			Rotifers	0.325			
			%WWTF	0.218			
			Temp	0.179			
2006 without %WWTF	May–August	Lower	Total_Flow_cfs	0.423	0.561	P<0.001	20
			NO2+NO3	0.371			
			Rotifers	0.325			
			Cl	0.226			
			Temp	0.179			
2007	June–September	Lower	%Natural_Flow	0.402	0.539	P<0.001	29
			%Flow_augmentation	0.368			
			SRP	0.289			
			SC	0.264			
2007 without SRP	June–September	Lower	%Natural_Flow	0.402	0.462	P<0.001	29
			%Flow_augmentation	0.368			
			SC	0.264			
			Rotifers	0.226			
			Total_Flow_cfs	0.188			
2008	May–August	Lower	%Natural_Flow	0.546	0.698	P<0.001	29
			Natural_Flow_cfs	0.544			
			SRP	0.425			
2008 without SRP	May–August	Lower	Natural_Flow_cfs	0.544	0.604	P<0.001	29
			%Natural_Flow	0.546			
			Copepods	0.354			

Table 12.   Average percent biovolumes of select algal taxa grouped according to chlorophyll-*a* and seasonal growth phase in the Tualatin River, Oregon.

[Chlorophyll-*a* (Chl-*a*) concentrations in micrograms per liter. **Abbreviations:** <, less than; >, greater than; sp., species]

Algal taxa	Algal division	Algal biomass condition, in relative biovolume (percent)				Average change in relative biovolume from bloom to late season (percent)
		Chl-*a* < 10 (Early eason)	Chl-*a* 10–20 (Growth phase)	Chl-*a* > 20 (Bloom)	Chl-*a* < 10 (Late season)	
Anabaena flos-aquae	Bluegreen	6	2	26	0	-26
Aphanizomenon flos-aquae	Bluegreen	1	1	6	0	-6
Stephanodiscus binderanus	Diatom	28	10	14	10	-4
Melosira ambigua	Diatom	0	0	2	0	-2
Rhopalodia gibba	Diatom	0	0	1	0	-1
Sphaerocystis schroeteri	Green	0	2	1	0	-1
Actinastrum hantzschii	Green	0	2	1	0	-1
Cryptomonas erosa	Cryptophyte	9	36	29	37	8
Cyclotella pseudostelligera	Diatom	4	5	1	5	4
Stephanodiscus hantzschii	Diatom	21	8	3	5	2
Fragilaria crotonensis	Diatom	0	0	0	2	2
Melosira distans alpigena	Diatom	1	1	0	2	2
Rhodomonas minuta	Cryptophyte	0	2	1	3	2
Cyclotella meneghiniana	Diatom	1	3	1	3	2
Chlamydomonas sp.	Green	1	8	4	6	2
Glenodinium sp.	Dinoflagellate	0	1	1	2	1
Melosira varians	Diatom	1	0	0	1	1
Gomphonema angustatum	Diatom	1	0	0	1	1
Achnanthes lanceolata	Diatom	2	0	0	1	1
Eudorina elegans	Green	2	0	0	1	1
Melosira granulata	Diatom	0	2	1	2	1
Synedra ulna	Diatom	4	1	0	1	1
Ankistrodesmus falcatus	Green	0	1	0	1	1
Scenedesmus quadricauda	Green	2	3	1	2	1

Table 13.   Pairwise analysis of similarity (ANOSIM) for phytoplankton samples collected each year by sequential sampling date, lower Tualatin River (river mile 11.5–3.4), Oregon.

[**Abbreviations:** Global R, Global Rho value; R, Rho value; P, probability]

Year	Pairwise significance			Global significance	
	Global R	Level	Pairwise tests	R statistic	Level
2006	0.851	P<0.001	06-26-06, 07-07-06	0.96	P<0.05
			07-07-06, 07-24-06	0.98	P<0.05
			07-24-06, 08-07-06	1.00	P<0.05
			08-07-06, 08-31-06	0.20	P<0.01
2007	0.736	P<0.001	06-27-07, 07-05-07	0.21	P<0.2
			07-05-07, 07-13-07	0.86	P<0.01
			07-13-07, 07-19-07	0.82	P<0.01
			07-19-07, 08-13-07	0.86	P<0.01
			08-13-07, 09-06-07	0.61	P<0.05
2008	0.737	P<0.001	06-17-08, 07-01-08	0.32	P<0.05
			07-01-08, 07-09-08	0.62	P<0.05
			07-09-08, 08-06-08	0.89	P<0.01
			08-06-08, 08-19-08	0.89	P<0.05

Bioassay Experiment Results

The bioassay experiments were designed to assess the potential effects of phosphorus concentrations and treated effluent from the Rock Creek WWTF on the plankton assemblages. Experimental results were evaluated by measuring changes in Chl-a (growth) and dissolved oxygen (photosynthetic activity) over the course of each incubation period. Although many experiments were conducted, and under a variety of river conditions, results presented here focus on the two earliest experiments conducted May 19 and June 11, 2008, before the Wapato event caused unusual plankton and water chemistry conditions, and on the June 30 experiment that was conducted at the onset of the Wapato event but under conditions where blue-green algae were not identified at the Rood Bridge site.

Phytoplankton populations in these experiments were dominated by diatoms (see fig. 21 for the specific dominants). Although some algal activity was noted during the first bioassay in May, the DO production was small and gains were just 2–6 percent of saturation in all light bottles owing to the low algal abundance. Gains in Chl-a and DO were smaller in the 50 percent wastewater sample, but the difference was not large. Similarly, addition of SRP had no clear or repeatable effect on either Chl-a or DO production in the May experiments. On June 11, the 30 percent effluent sample produced a 12 percent increase in Chl-a compared with just 2.9 percent for the 0 percent effluent sample; Chl-a concentrations in the 50 percent sample declined 15 percent. On June 30, Chl-a gains of 44–58 percent were observed over the course of the experiment, with the smallest gains occurring in the 50 percent effluent treatment (fig. 21). Any conclusions from these initial bioassays should be tempered by the limited range of conditions tested and the lack of sufficient replicate samples. Regardless, these results suggest that, at least under certain conditions, effluent concentrations of 30 percent may stimulate algal growth, whereas 50 percent may either inhibit (June 11) or stimulate growth to a lesser degree (June 30). The subject bears further study because of the large fraction of treated effluent present in the Tualatin River during summer, but these bioassays did not show a repeatable positive or negative effect of treated effluent or added phosphorus on algal growth or photosynthetic activity.

Declines in Phytoplankton Populations

A notable decline in phytoplankton abundance in the Tualatin River in late July and August began in 2002, after the 2001 drought year, and continued from 2003 to 2009, resulting in lower DO concentrations and more frequent occurrence of DO standard violations (fig. 8). Although Chl-a levels upstream of the reservoir reach of the Tualatin River are generally low (less than 6 µg/L), the size of the population entering that reach appears to be a key determinant of population levels downstream. As lower concentrations of algae come into the reservoir reach, the blooms do not reach abundances as high as in years past.

The cause of this decline in Chl-a is unknown, but several factors could contribute, as suggested by the BEST analyses. Increased reservoir releases in July for flow augmentation, for example, coincide with these trends in decreased upstream algae populations (fig. 22). Another factor that will require more study, and which might be correlated with increased reservoir releases, is the possibility of slight increases in minumum turbidity levels (figs. 23, 24, and 25). Increased turbidity during the summer of 1996 was known to be an important factor holding down algal populations that year (fig. 25), suggesting that algal production in the Tualatin River is susceptible to light limitation. Water-quality modeling supports that conclusion (Rounds and others, 1999). No observable trend in water temperature was found to explain lower algal abundances in recent years, based on analysis of continuous data from RM 24.5 for the month of August going back to 1997.

A reduction in the upstream algal seed source, or "inoculum," has important implications because the maximum algal biomass in the lower river is strongly dictated by its initial population size entering the reservoir reach, which is now considerably lower at Rood Bridge, RM 38.4 (figs. 7 and 23). Even small reductions in the inoculum could translate into reductions at downstream sites. Ratios of measured Chl-a at Elsner to those upstream at Rood Bridge (here termed the "chlorophyll-a growth ratio") are useful in evaluating algal growth in this reach of the river over time. Historically, this reach was characterized by a rapidly increasing algal population, but that has typically not been the case in recent years. In the past, the population at Elsner was often 10 times larger than at Rood Bridge, whereas growth ratios in this reach in recent years are much smaller, and often less than 1, indicating times such as August 2004 when the Chl-a actually declined in this reach (fig. 26).

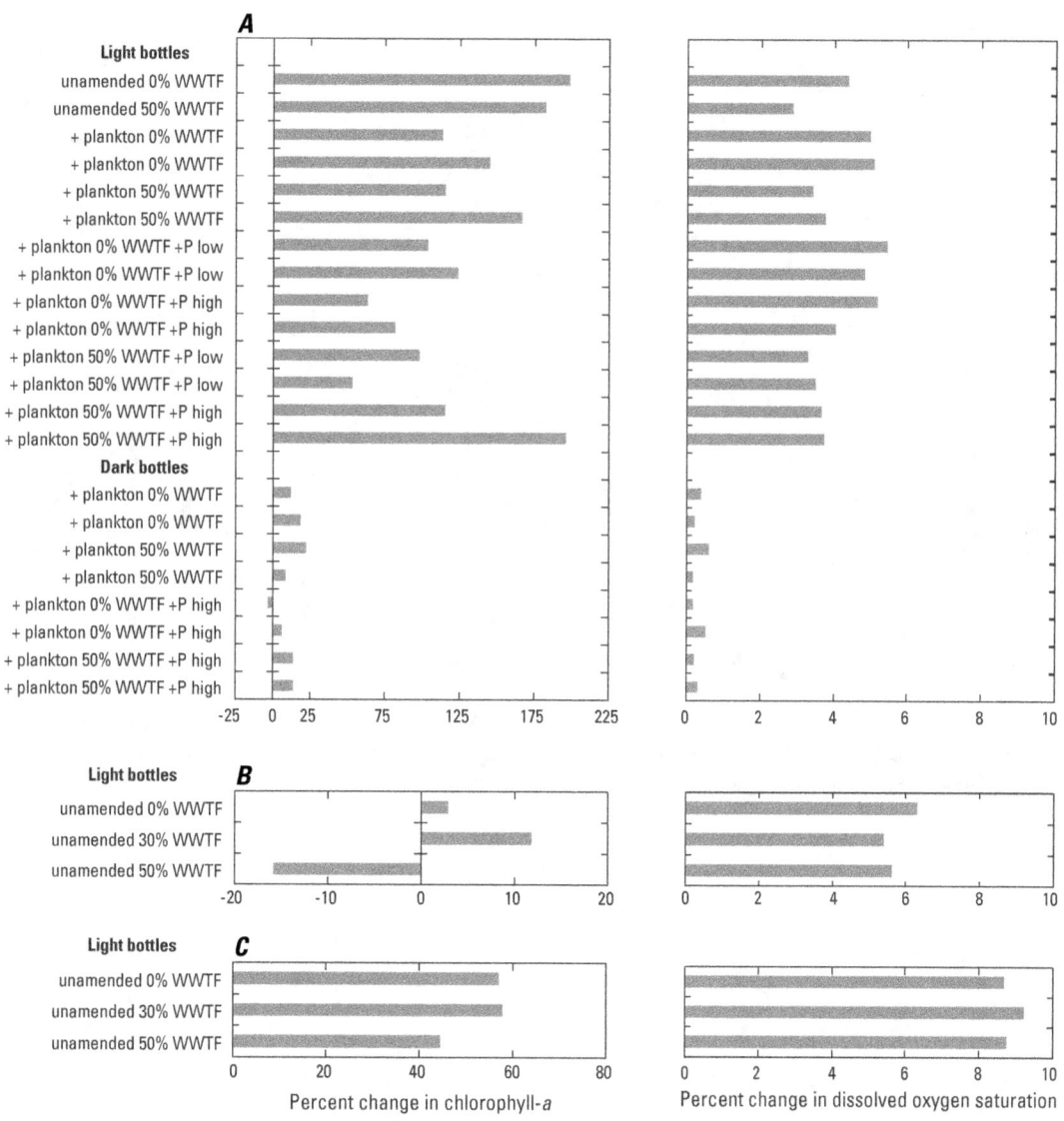

Figure 21. Changes in chlorophyll-*a* and dissolved oxygen concentrations during bioassay experiments on (*A*) May 19, (*B*) June 11, and (*C*) June 30, 2008, Tualatin River, Oregon. On May 19, unamended samples were dominated by *Synedra*, *Stephanodiscus*, *Cymbella*, and other diatoms; amended samples dominated by *Asterionella*, *Stephanodiscus*, and *Melosira*. On June 11, unamended samples dominated by *Anabaena* and *Stephanodiscus*. On June 30, *Cryptomonas erosa*, *Eunotia pectinalis*, and *Melosira varians* dominated unamended samples. Abbreviations: WWTF, wastewater treatment facility effluent; P, phosphorus; %, percent; DO, dissolved oxygen.

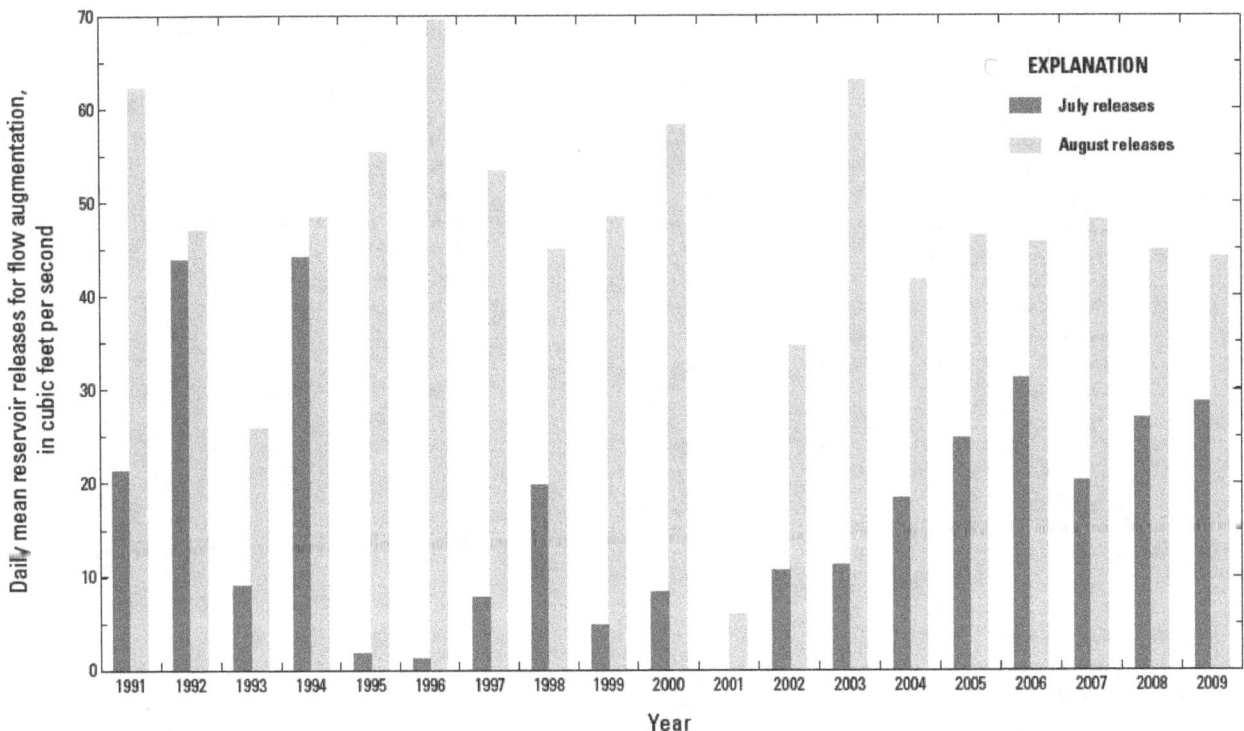

Figure 22. Time series of water releases from Barney Reservoir and Hagg Lake, Oregon, by Clean Water Services for flow augmentation, July and August, 1991–2009. Data from Clean Water Services.

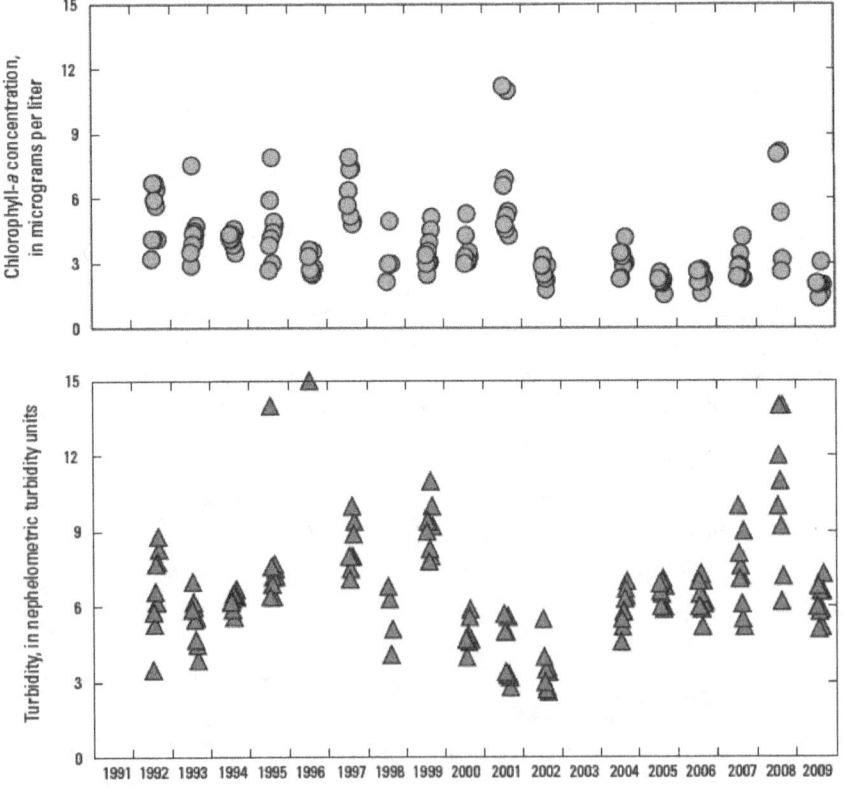

Figure 23. Time series of (*A*) chlorophyll-*a* concentration and (*B*) turbidity in the Tualatin River at Rood Bridge (river mile 38.4), Oregon, July–August 1992–2009. Data from Clean Water Services.

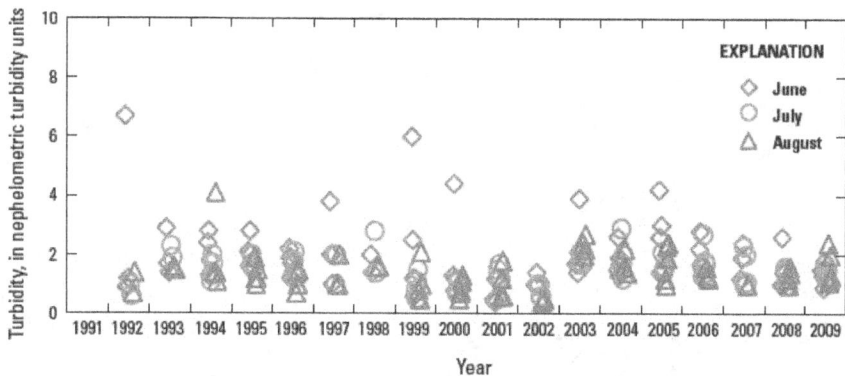

Figure 24. Time series of turbidity in the Tualatin River at Cherry Grove (river mile 67.8), Oregon, June–August 1991–2009. Data from Clean Water Services.

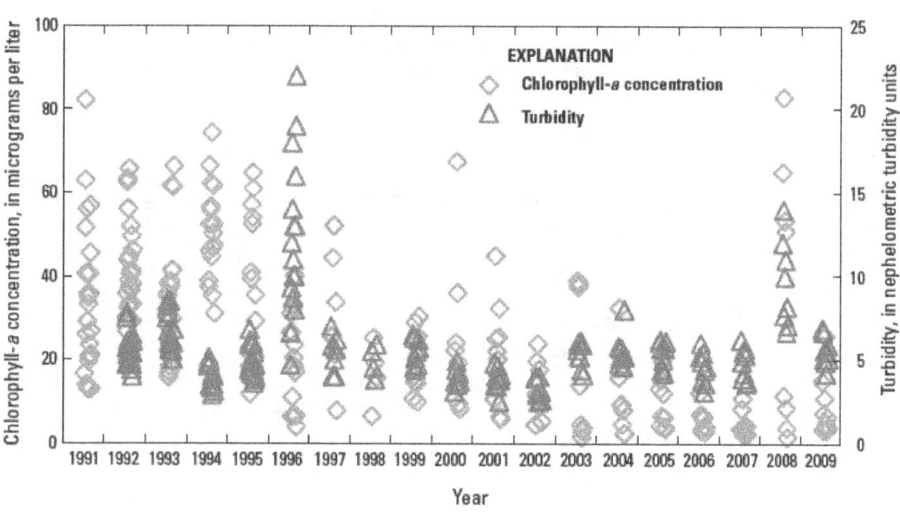

Figure 25. Time series of chlorophyll-*a* concentrations and turbidity in the Tualatin River at Elsner Road (river mile 16.2), Oregon, July and August 1991–2009. Data from Clean Water Services.

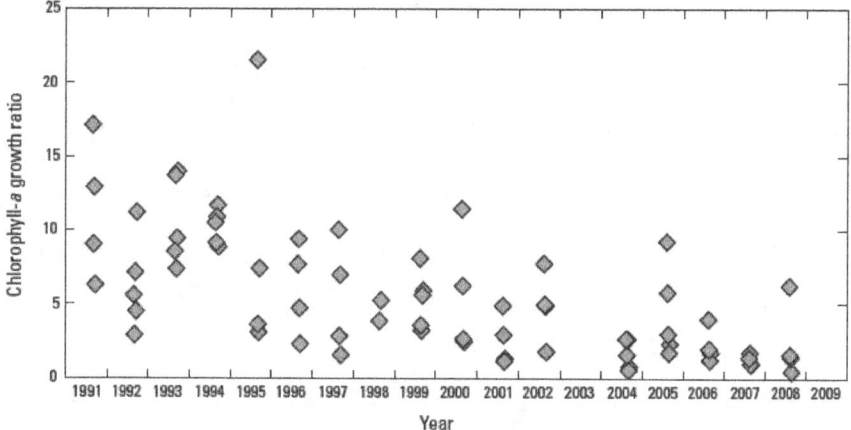

Figure 26. Trend in the chlorophyll-*a* growth ratio in the reach between Rood Bridge (river mile [RM] 38.4) and Elsner Road (RM 16.2), Tualatin River, Oregon, during August, 1991–2008. The growth ratio is the chlorophyll-*a* concentration at Rood Bridge divided by the chlorophyll-*a* concentration at Elsner Road. Data from Clean Water Services.

Declines in the Chl-*a* growth ratios in the Rood to Elsner reach suggest that something is affecting algal growth or accumulation of biomass. Although it is possible that factors such as flow, nutrient levels, or the percentage of flow from reservoir releases and WWTFs also contribute to the declines, higher turbidity may be able to explain the reduced algae levels because algae require light for photosynthesis. Increased turbidity levels, accompanied by a reduced upstream inoculum, might combine to limit algal growth in the reservoir reach. Historical July–August turbidity and Chl-*a* concentrations at Elsner are plotted in figure 27. No clear threshold of turbidity is indicated in that plot as a necessary condition to limit algal growth. Some of the scatter in that plot is caused by higher turbidity at higher algal abundance, which is expected; regardless, the bulk of the highest Chl-*a* concentrations occur when the turbidity values are less than about 7 nephelometric turbidity units (NTUs), thus offering more evidence suggestive of light limitation on algal growth. Apparent increases in summer turbidity minima in recent years at Rood Bridge (fig. 23) and Elsner (fig. 25), at around 4–8 NTUs, may be high enough to begin exerting a growth limitation that, along with other factors, is suppressing algal populations in July and August.

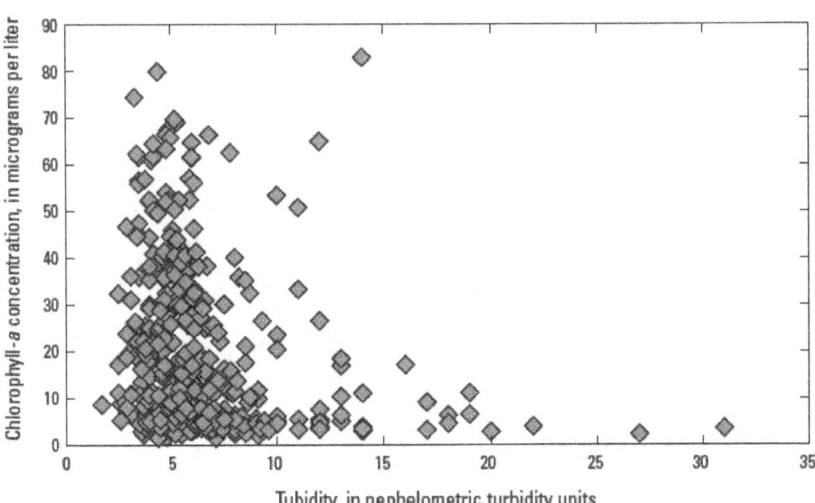

Figure 27.   Relation between turbidity and chlorophyll-*a* concentrations in the Tualatin River at Elsner Road (river mile 16.2), Oregon, May–August 1992–2009. Data from Clean Water Services.

Changes in River Hydrology and Wastewater Management, 1991–2008

To understand why algal populations have generally declined in July and August since 2003, it is important to quantify and assess the changes in hydrology and improvements in wastewater treatment that have occurred over the past two decades, as those factors have an influence on algal populations. Nutrient reductions from the WWTFs, especially lower SRP levels, have greatly reduced the magnitude and duration of algal blooms, by design. Moreover, water releases during summer, mostly from Hagg Lake, and now also from Barney Reservoir, greatly increase streamflow during critical low-flow periods. In the upper reaches of the Tualatin River at Cherry Grove, flows in August have increased (fig. 28) owing to releases from Barney Reservoir. Increased summer flow augmentation is especially evident in the lower river at West Linn (fig. 29), where the frequency and duration of lower flows (darker blue colors) have decreased over time, while the frequency of higher flows (lighter blue colors) has increased in summer, especially in August. The particularly low flows at West Linn in 1991–95 were partly due to large (50 ft³/s) withdrawals at the Oswego Canal (RM 6.7), which decreased to 10–15 ft³/s during 1996–2003 and 1 ft³/s or less thereafter. Close examination of the data reveals short pulses of higher flows in August (Julian days 213 to 243), indicated by the red bars in the black box of figure 29. Flow augmentation (and decreased withdrawals) has essentially raised the baseflow condition for the river (fig. 30). By shortening the residence time and reducing phosphorus concentrations, the direct consequence is a decrease in algal populations in July and August.

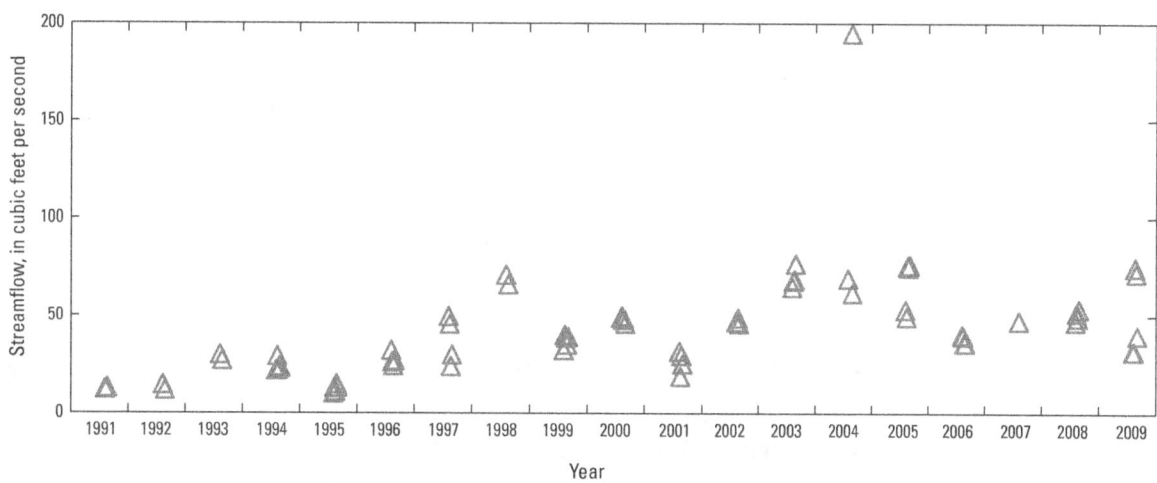

Figure 28.   Time series of streamflow in the Tualatin River at Cherry Grove (river mile 67.8), Oregon, August 1991–2009. Data from Clean Water Services.

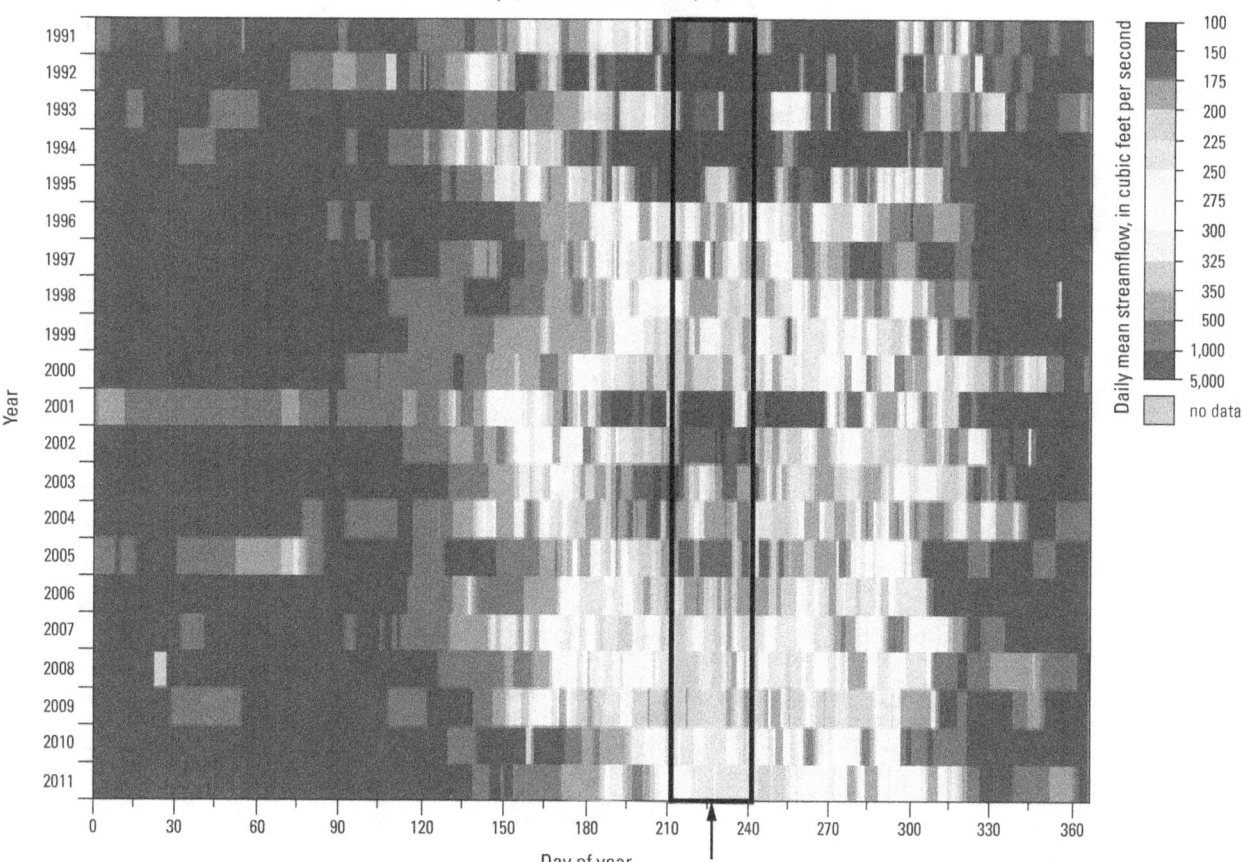

Figure 29.   Time series of streamflow in the Tualatin River at West Linn (river mile 1.8), Oregon, 1991–2011. Black box indicated by arrow is the month of August.

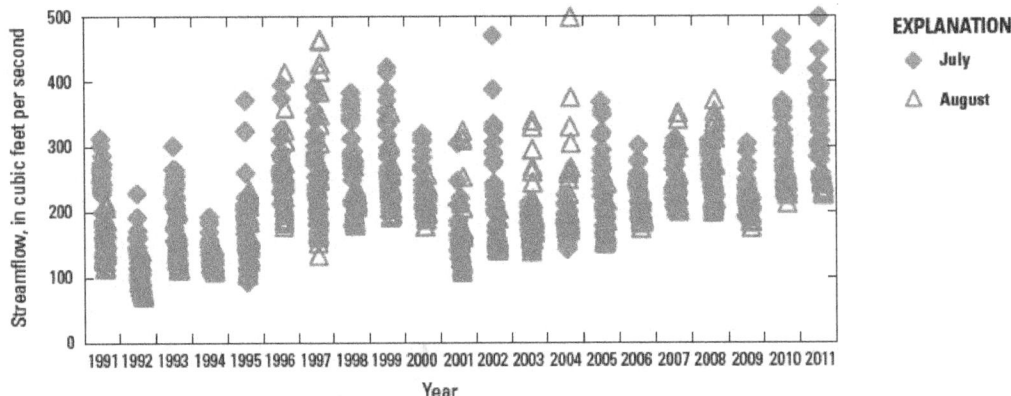

Figure 30.   Time series of streamflow in the Tualatin River at West Linn (river mile 1.8), Oregon, July and August, 1991–2011.

Higher flows in the river also result from higher discharges of treated effluent from the Rock Creek WWTF (fig. 13), which serves a population that has grown faster than the population in the lower basin, which was developed earlier and is served by the Durham WWTF. Greater percentages of WWTF effluent and flow augmentation releases from upstream reservoirs result in decreased percentages of natural flow in the river in summer. This is important because the natural flow contains a greater amount of both Chl-*a* and diatoms (fig. 31). Conversely, flow augmentation and WWTF effluent as a percentage of the total flow were negatively correlated with these algal indicators at the RM 24.5 and 38.4 sites.

Although the importance of Chl-*a* in maintaining minimum levels of DO in the Tualatin River is well established (Rounds and others, 1999), the types of algae occurring in the river also have an influence on the health of this ecosystem. Diatoms and other algae make important contributions to the river's food web, supporting seasonal populations of zooplankton, which, in turn, are food for planktivorous fish. From a management standpoint, because of their high fatty acid content, diatoms are nutritious and much preferred over blue-green algae (Caramujo and others, 2008), which can form surface blooms, and sometimes produce harmful toxins. Identification of potential management options that contribute to conditions supportive of diatoms might also lessen competition from blue-green algae and the problems associated with their blooms.

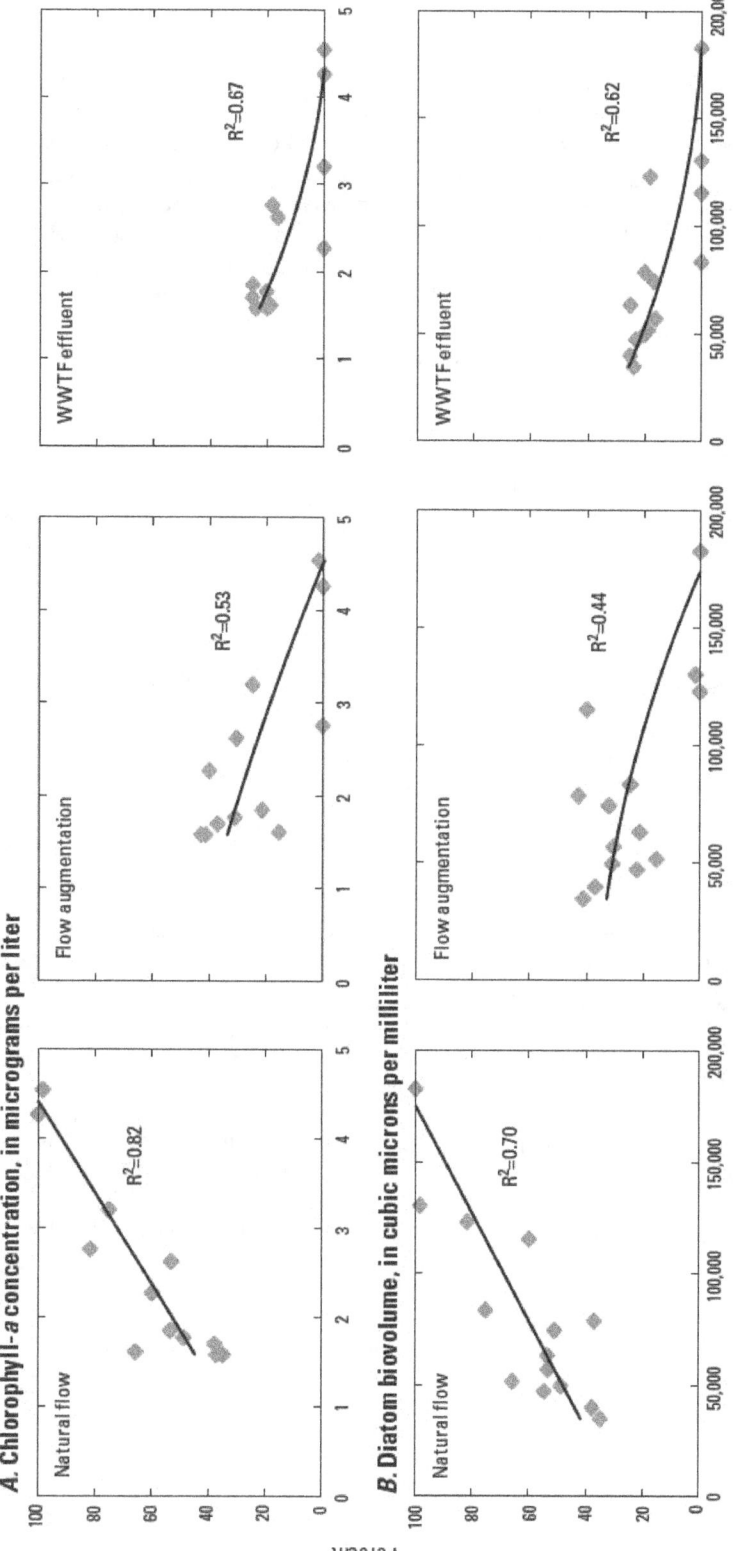

Figure 31. Relations between the percentage of natural flow, flow augmentation, and wastewater treatment facility (WWTF) effluent and (*A*) chlorophyll-*a* and (*B*) diatom biovolume in the Tualatin River, Oregon, at river miles 38.4 and 24.5. Two outlier samples collected during blooms in July 2007 and July 2008 with chlorophyll-*a* concentrations greater than 10 micrograms per liter were not included.

Case Studies of Low Dissolved Oxygen Events and Bloom Crashes in 2003–08

Although phytoplankton blooms still occur, they tend to be shorter in duration and terminate earlier than in years past. In previous years, Chl-*a* concentrations would remain high enough to sustain the DO throughout most or all of summer, at least through late August; in recent years, while blooms sometimes decline gradually, they often crash abruptly in mid-summer. This earlier onset of algal declines often coincides with peak water temperatures of 24–25°C in late July or early August, which can contribute to low DO conditions because DO solubility decreases with increasing temperature and sediment oxygen demand (SOD) increases with increasing temperature. To augment the general analysis of algal declines in July-August and to better determine the factors leading to algal declines, an examination of specific algal bloom-decline events during 2003–08 was undertaken.

Low Dissolved Oxygen Events

Daily minimum DO measurements at the Oswego Dam in 2003–08 show 9 periods, labeled A to I in figure 32, when concentrations were less than 6.5 mg/L, the State standard based on the 30-day mean DO concentration. The occurrence of low-DO conditions during summer was directly related to declines in algal populations and Chl-*a* levels. Specific timeframes, duration, flow, light, and turbidity conditions, and possible causes of each of the events are listed in table 14. Although short-term declines in DO may occur during summer in response to (1) clouds and rain that temporarily disrupt photosynthesis or shorten the residence time via streamflow increases, or (2) rare discharges of high ammonia levels from the WWTFs, the longest duration low-DO conditions always occurred after the decline of an algal bloom. During these times, the daily minimum DO concentrations often were less than 6.5 mg/L for as many as 77 days (table 14 and fig. 4), and sometimes less than the absolute minimum State standard of 4 mg/L for shorter durations.

Table 14. Low dissolved oxygen events in the lower Tualatin River at the Oswego Dam, Oregon, 2003–08, and potential contributing factors.

[**Low dissolved oxygen event:** See figure 32. Low dissolved oxygen (DO) events resulting in DO concentrations less than 6.5 mg/L. The 6.5-mg/L State standard is based on the 30-day mean concentration. **Abbreviations:** mg/L, milligrams per liter; WWTF, wastewater treatment facility; ft^3/s, cubic feet per second]

Low dissolved oxygen event	Year	Dates when dissolved oxygen was less than 6.5 mg/L	Approximate duration of daily minimum dissolved oxygen less than 6.5 mg/L (days)	Minimum dissolved oxygen	Potential cause / Contributing factors
A	2003	July 11–November 1	77, intermittent	3.8 mg/L on August 10	Algal population decline; high water temperature
B	2004	July 26–Sept 11	48	4.3 mg/L on August 23	Algal population decline; high water temperature
C	2005	August 8–Sept 22	46	4.6 mg/L on August 30	Algal population decline
D	2006	May 22–26	4	4.8 mg/L on May 23	0.87 inch of rain on May 21 preceded this event
E	2006	July 29–Oct 7	53, intermittent	5.4 mg/L on July 30	Algal population decline; high water temperature
F	2007	July 19–Sept 22	57, intermittent	4.7 mg/L on July 23	Low solar radiation and 0.5 inch of rain; high water temperature
G	2008	May 21–26	6	6.0 mg/L on May 26	Ammonia release from the Rock Creek WWTF; multi-day 1-inch rain event
H	2008	July 1–9	9	5.2 mg/L on July 5	Algal decline; 0.4 inch of rain 2 days prior
I	2008	August 12–October 12	60	3.8 mg/L on August 20	Crash of large algal population; increase in flow augmentation up to 125 ft^3/s; zooplankton grazing; rainfall

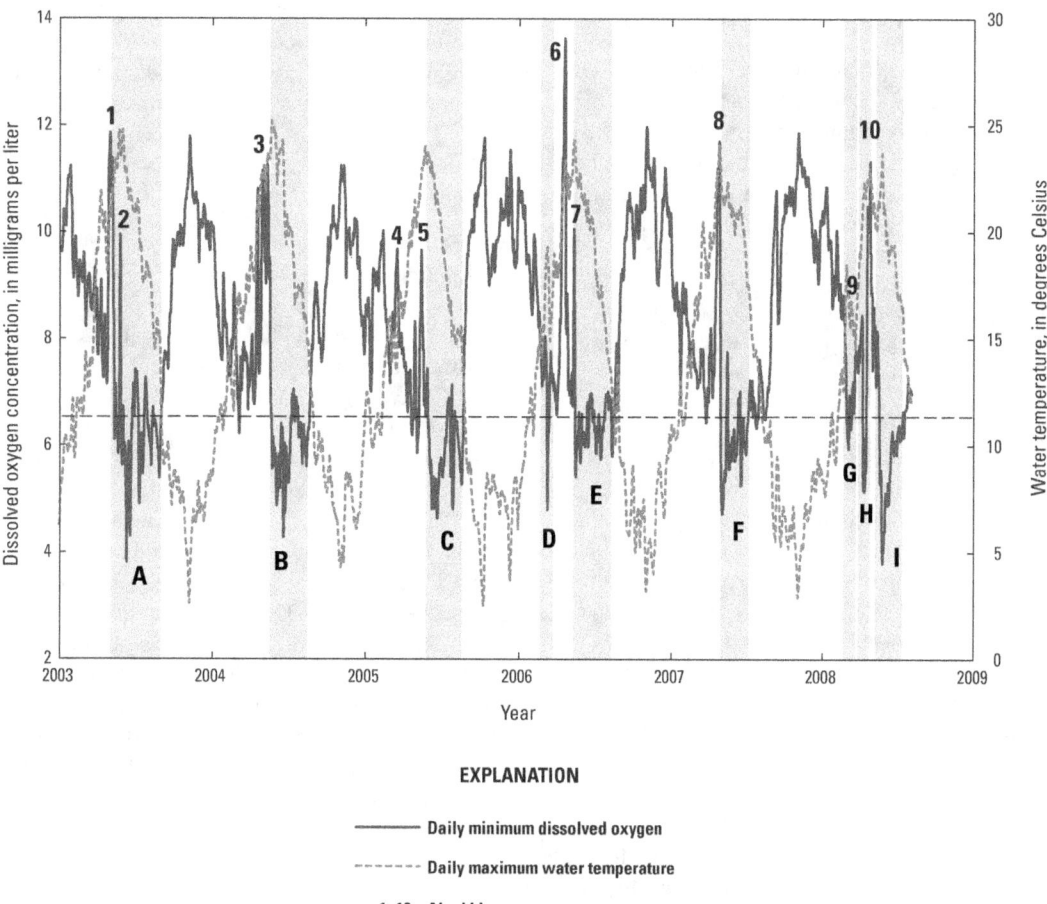

EXPLANATION

———— Daily minimum dissolved oxygen

- - - - - Daily maximum water temperature

1–10   Algal blooms

A–I   Low dissolved oxygen events

Figure 32.   Dissolved oxygen (DO) concentrations and water temperature in the Tualatin River at Oswego Dam (river mile 3.4), Oregon, 2004–08, highlighting algal blooms and low-DO events. Shaded areas indicate periods when DO was less than 6.5 milligrams per liter. See tables 14 and 15 for information regarding the conditions leading up to the low-DO event and potential contributing factors in bloom declines.

Specific Bloom Crashes

A combination of algae, water-quality, weather, and flow data were analyzed for 2003–08 to determine some of the factors that may have triggered the 10 bloom crashes listed in table 15. For this analysis, data from the continuous monitor at the Oswego Dam, CWS discrete data for the mainstem, and qualitative (in 2005) and quantitative (in 2006–08) plankton conditions were utilized. Total photosynthetically active radiation (PAR) and rainfall data from the Durham WWTF monitoring site also provided weather data to help discern whether clouds or rain played a role in any bloom crashes.

2003

In 2003, algal growth started in late May, leading to a bloom that peaked in late June and early July. No species composition data are available to determine the type of algae that caused a 70 µg/L peak in Chl-*a* concentration at Stafford Road (RM 5.5). Although low SRP concentrations at Boones Ferry (0.01 mg/L, RM 8.7) may have contributed to or triggered the bloom crash in the second week of July, cloudy conditions also may have contributed (fig. 33 and table 15). The low SRP concentrations probably resulted from the combined effect of lower than normal concentrations of SRP in treated effluent being discharged from the Durham WWTF (0.012 mg/L on July 13–14), and uptake of SRP by phytoplankton, which had peaked at 64 µg Chl-*a*/L the week before at Boones Ferry, about a half mile downstream from the Durham WWTF (fig. 2), but declined to 9 and 4 µg Chl-*a*/L, respectively, on the next two samplings.

A second bloom in 2003 peaked on July 25, followed by a second crash (fig. 33) that also occurred when SRP levels at Elsner were quite low (0.01 mg/L) and turbidity increased slightly. For 77 days after the crash, the daily minimum DO was intermittently less than 6.5 mg/L at the Oswego Dam, reaching a minimum DO of 3.8 mg/L on Aug 10 (table 14). Chl-*a* levels rebounded slightly to 20 µg/L on August 1, but a rainfall event prevented further growth by increasing flow and decreasing both available light and the residence time (fig. 33).

2004

In 2004, a bloom peaked at 60 µg Chl-*a*/L on July 7 at Stafford Road, decreasing the SRP to 0.01 mg/L at Boones Ferry. The bloom declined slowly thereafter, rebounded at times, but finally crashed on July 22 (table 14). The cause of this crash appears to be due to limitation by phosphorus as well as flow factors, as this was about the time when the total percentage of WWTF effluent reached its seasonal peak. From July 15 to July 22, the percentage of WWTF effluent in the river at Boones Ferry increased from about 41 to 48 percent.

Whatever the cause, for 48 days from July 26 to September 11, the daily minimum DO was below 6.5 mg/L. The lowest DO value of 4.3 mg/L was observed on Aug 23, when the maximum water temperature was about 23°C (table 14).

2005

A small and steady population of *Aulacoseira* developed in June 2005, but was interrupted twice by two approximately 0.35-in. rain events in late June and early July. The first rain event produced a streamflow at West Linn of 453 ft³/s (greater than the 300 ft³/s level below which algal growth can typically be sustained) and a turbidity of 16 NTU. These conditions were followed by a moderate population of blue-green algae (*Aphanizomenon* and *Anabaena*) that developed in the lower river at Elsner (RM 16.2) and produced the peak Chl-*a* concentration of the season, 40 µg/L at the Oswego Dam on July 18. Thereafter, algal biomass was lower and declined slowly to less than 10 µg/L at the end of July. Zooplankton, primarily cladocerans, were observed in samples from Boones Ferry to the Oswego Dam at this time, suggesting the possibility that grazing may have played a role in the algal biomass decline. The lack of algal photosynthesis was followed by a decline in the DO to a minimum concentration of 4.6 mg/L on August 30. The 6.5-mg/L DO State standard (based on the 30-day mean concentration) was not met for 72 consecutive days, from August 12 to October 22.

2006

In 2006, an early July bloom was followed by a rapid crash; the population rebounded only to decline again at the end of July. The first crash coincided with a period of cloudy weather, but the algae rebounded once light conditions improved. At the end of July, this second bloom declined, coinciding with a small decline in PAR and a slight increase in turbidity at the Oswego Dam of 9 formazin nephelometric units (FNUs) (fig. 34). The daily minimum DO concentration was less than 6.5 mg/L at the Oswego Dam for 53 of the 71 days from July 29 to October 7, reaching a minimum of 5.4 mg/L on July 30 (table 14).

A marked shift in the phytoplankton species composition also occurred during this time period in 2006: the biovolume of diatoms, mostly *Stephanodiscus binderanus*, declined from 70 to less than 15 percent of the total biovolume, whereas the biovolume of Cryptophytes, mostly *Cryptomonas erosa* and *Rhodomonas minuta*, increased from less than 25 to almost 80 percent from July 7 to August 7 (fig. 34). This change coincided with an increase in releases from Hagg Lake, and a further decrease in the percentage of natural flow, as well as higher populations of zooplankton, all of which seem to be correlated with algal declines.

Table 15. Algal bloom–crash sequences in the Tualatin River, Oregon, during 2003–08, and possible contributing factors.

[Algal blooms: See figure 32. Favorable conditions for phytoplankton growth defined as: greater than 1,250 (µE/m²)/s daily maximum photosynthetically active radiation at the Durham WWTF, daily median turbidity values less than 7 FNU at the Oswego Dam, SRP greater than or equal to 0.015 mg/L, and streamflow less than 300 ft³/s at West Linn. Water temperature (daily maximum) from the Oswego Dam site. Abbreviations: (µE/m²)/s, microeinsteins per square meter per second; WWTF, wastewater treatment facility; FNU, formazin nephelometric units; SRP, soluble reactive phosphorus; mg/L, milligrams per liter; ft³/s, cubic feet per second; µg/L, micrograms per liter; NTU, nephelometric turbidity units; DO, dissolved oxygen; m³, cubic meters]

| Algal blooms | Year | Algal conditions | Peak chlorophyll-*a*, location | Growing conditions for phytoplankton in lower river | | | | | Streamflow at West Linn during crash | | Possible trigger or contributing factors |
				Solar radiation	Water temperature	Turbidity	SRP	Streamflow	Daily mean flow on first day	Average during crash	
1	2003	Growth started in late May; first bloom peaks on June 30, then crashes from July 7–13	70 µg/L, Stafford Road	**Unfavorable**	23–25 C	Favorable	**Unfavorable, 0.01 mg/L**	Favorable	183	187	Cloudy weather conditions; low SRP (0.01 mg/L) at Boones Ferry on June 30
2	2003	Second bloom peaks on July 25, then crashes from July 26–29	50 µg/L, Oswego Dam	**Unfavorable**	22–23 C	**Unfavorable, 6–9 NTU**	**Unfavorable, 0.01 mg/L**	Sometimes favorable	150	150	Cloudy weather conditions; Somewhat higher turbidities in the lower river may have limited light availability; Low SRP (0.01 mg/L) at Elsner; subsequent growth limited by higher flows and less inocula coming into the reservoir reach from upstream
3	2004	Algal bloom peaked about July 7, then declined for 9 days with minor rebounds; final crash on July 22	60 µg/L, Stafford Road	Some clouds, but mostly favorable	23–25 C	Favorable	**Unfavorable, 0.01 mg/L**	Favorable	211	183	Low SRP (0.01 mg/L) at Boones Ferry on July 6; 0.012 mg/L at Stafford Road on July 12; 10 and 19 percent increase in WWTF discharge from the Rock Creek and Durham WWTFs increases the proportion of WWTF to 48 percent of the total flow
4	2005	Small population of *Aulacoseira* develops in June	~10 µg/L, Oswego Dam	**Unfavorable**	20–21 C	**Briefly unfavorable, up to 16 NTU**	Favorable	**Unfavorable**	340	407	0.35-inch rain event interrupted the first bloom
5	2005	Blooms of *Aulacoseira*, then mixed with *Aphanizomenon* and *Anabaena* peaks on July 18, and then declines	40 µg/L, Oswego Dam	Favorable	24 C	Favorable	Favorable	Favorable	173	184	Zooplankton (cladocerans and copepods) abundant [not quantified], so crash may have been due to grazing losses
6	2006	*Cryptomonas erosa*, *Chlamydomonas* sp., and several diatoms (*Aulacoseira* and *Stephanodiscus*) dominate assemblage; bloom peaks on July 3, declines, then rebounds	80 µg/L, Stafford Road	**Unfavorable**	24 C	Favorable	**Unfavorable, 0.011 mg/L**	Favorable	197	217	Low light from cloudy conditions; Low SRP (0.011 mg/L) at Stafford on July 3; flow augmentation increases to 98 ft³/s, zooplankton (mostly cladocerans) peak at about 15,000 per m³

Table 15. Algal bloom–crash sequences in the Tualatin River, Oregon, during 2003–08, and possible contributing factors.—Continued

[Algal blooms: See figure 32. Favorable conditions for phytoplankton growth defined as: greater than 1,250 (μE/m^2)/s daily maximum photosynthetically active radiation at the Durham WWTF, daily median turbidity values less than 7 FNU at the Oswego Dam, SRP greater than or equal to 0.015 mg/L, and streamflow less than 300 ft^3/s at West Linn. Water temperature (daily maximum) from the Oswego Dam site. Abbreviations: (μE/m^2)/s, microeinsteins per square meter per second; WWTF, wastewater treatment facility; FNU, formazin nephelometric units; SRP, soluble reactive phosphorus; mg/L, milligrams per liter; ft^3/s, cubic feet per second; μg/L, micrograms per liter; NTU, nephelometric turbidity units; DO, dissolved oxygen; m^3, cubic meters]

Algal blooms	Year	Algal conditions	Peak chlorophyll-a, location	Growing conditions for phytoplankton in lower river					Streamflow at West Linn during crash		Possible trigger or contributing factors
				Solar radiation	Water temperature	Turbidity	SRP	Streamflow	Daily mean flow on first day	Average during crash	
7	2006	A second bloom of flagellates peaks on July 24, then declines	80 µg/L, Stafford Road	Favorable	24 C	Possibly unfavorable, up to 9 FNUs	Unfavorable, 0.011 mg/L	Favorable	224	210	Low SRP (0.011 mg/L) on July 24; Flow augmentation increases to 122 ft³/s; natural flow makes up less than 50 percent of total starting on July 20; Low light on July 30; cladoceran density 12,000 per m³ at Oswego Dam on July 24
8	2007	Stephanodiscus and Cryptomonas erosa dominate assemblage, which peaks on July 16 then crashes	68 µg/L, Stafford Road	Unfavorable	22–24 C	Favorable	Unfavorable, 0.005 mg/L	Sometimes favorable	224	261	0.5-inch rain event; Low SRP at Stafford Road (0.005 mg/L); Increase in flow augmentation up to 96 ft³/s; high zooplankton densities (25,000–50,000 cladocerans per m³ at Stafford and Oswego Dam) suggest grazing may have contributed to the algal decline
9	2008	Stephanodiscus dominates algal assemblage, which peaks on June 27, then declines	20 µg/L, Oswego Dam	Unfavorable	22 C	Favorable	Favorable	Favorable	257	258	0.4-inch rain event
10	2008	Anabaena, Aphanizomenon, and Cryptomonas erosa blooms from July 10 to August 4, then declines by August 11	50 µg/L on July 10 at Oswego Dam; 83 µg/L on July 24 at Elsner	Favorable	22–23 C	Favorable	Favorable	Favorable	250	275	Inocula ended; Increase in flow augmentation up to 125 ft³/s; high density of zooplankton (88,000 per m³ downstream from Elsner on July 16) suggests that grazing losses might also have played a role in the crash or in keeping the bloom from rebounding.

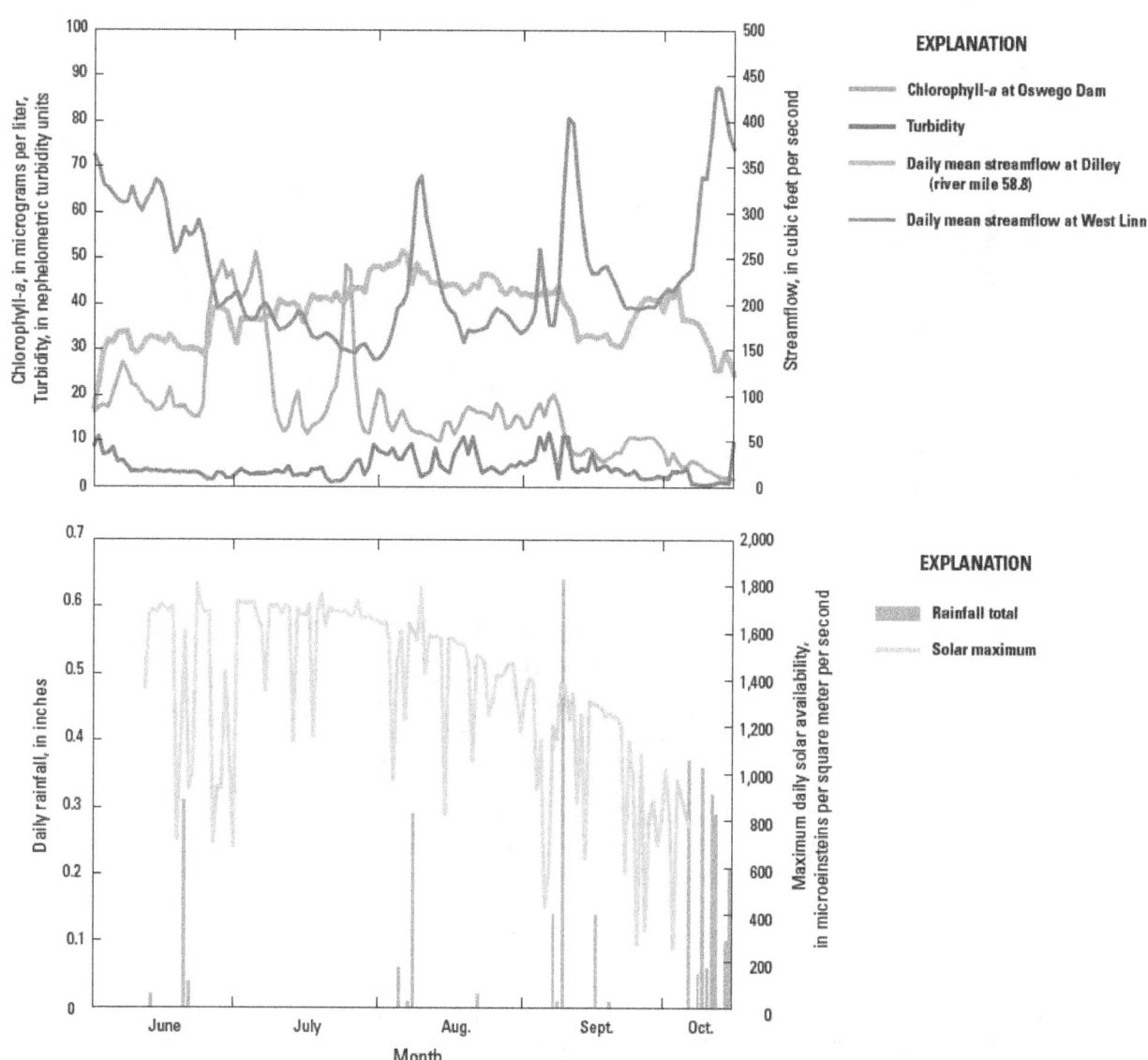

Figure 33. Time series of (A) chlorophyll-a, turbidity, and streamflow in the Tualatin River and (B) daily rainfall and solar maximum at the Durham wastewater treatment facility, Oregon, June–October 2003.

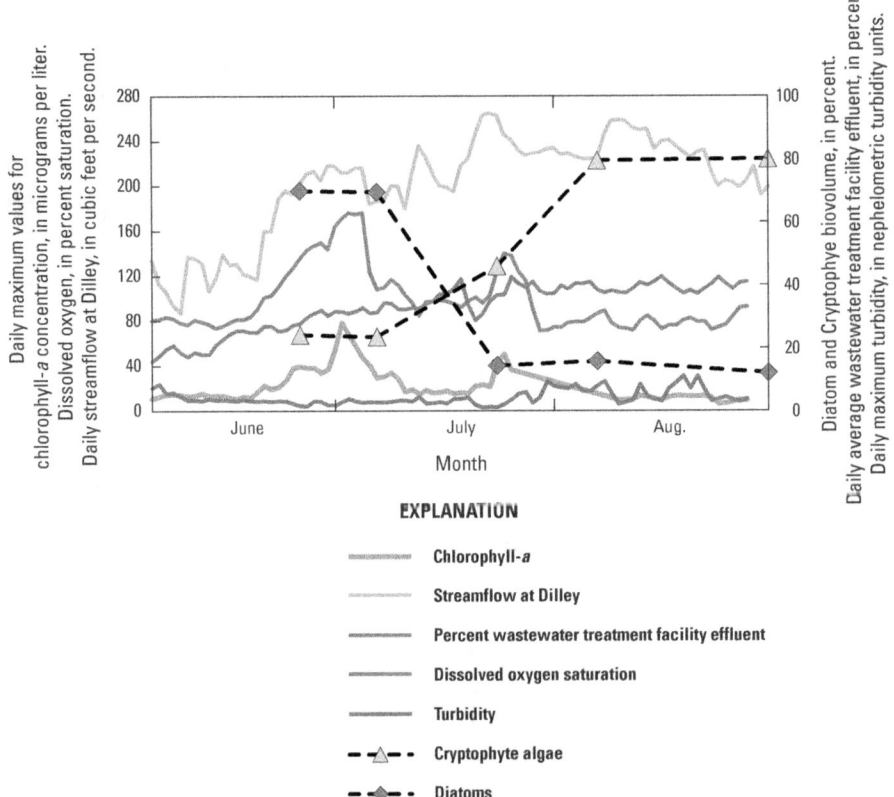

Figure 34. Time series of chlorophyll-*a*, streamflow, and water-quality conditions in the Tualatin River, Oregon, June–August 2006. The percent biovolume of diatoms and Cryptophyte algae at Stafford Road also are indicated. Chlorophyll-*a*, dissolved oxygen, and turbidity data from the continuous monitor at the Oswego Dam, RM 3.4.

2007

In 2007, persistent cloudy periods and rain resulted in unfavorable conditions for phytoplankton populations in late June and mid-July. A single bloom occurred, peaking in mid-July (fig. 35A) and crashing a week later from more rain and cloudy weather (fig. 35B). Although small rebounds in Chl-*a* occurred later in the summer, more clouds and rain again reduced algal biomass. The largest crash coincided with a pulse of increased flow from 209 to 312 ft³/s on July 20 at West Linn, a level that is higher than the general 300 ft³/s rule-of-thumb flow threshold above which the residence time tends to be insufficient to grow an appreciable algal population. The daily minimum DO was less than 6.5 mg/L at the Oswego Dam for 57 of the 66 days between July 19 and September 22, declining to 4.7 mg/L on July 23 (table 14).

2008

The first of three low-DO events in 2008 occurred in late May (see event G, fig. 32), when an ammonia release occurred after an interruption of the nitrification process at the Rock Creek WWTF and treated effluent concentrations were approximately 11 mg/L. Clouds and rain for several days also contributed to this low-DO event, which lasted for 6 days and produced a minimum DO concentration of 6.0 mg/L on May 26 (table 14).

In late June, a population of *Stephanodiscus* developed in the lower river. After reaching about 20 µg Chl-*a*/L at the Oswego Dam, unfavorable growing conditions caused by rain and cloudy weather caused the bloom to decline from June 29 into the first week of July, and Chl-*a* declined to 7 µg/L (fig. 36). The decline in Chl-*a* precipitated another low-DO event (daily minimum DO less than 6.5 mg/L) that started on July 1 and lasted 9 days, reaching a minimum DO of 5.2 mg/L on July 5 (table 14). By this time, the Wapato Lake blue-green algae bloom in the upper river had already started and was moving downstream.

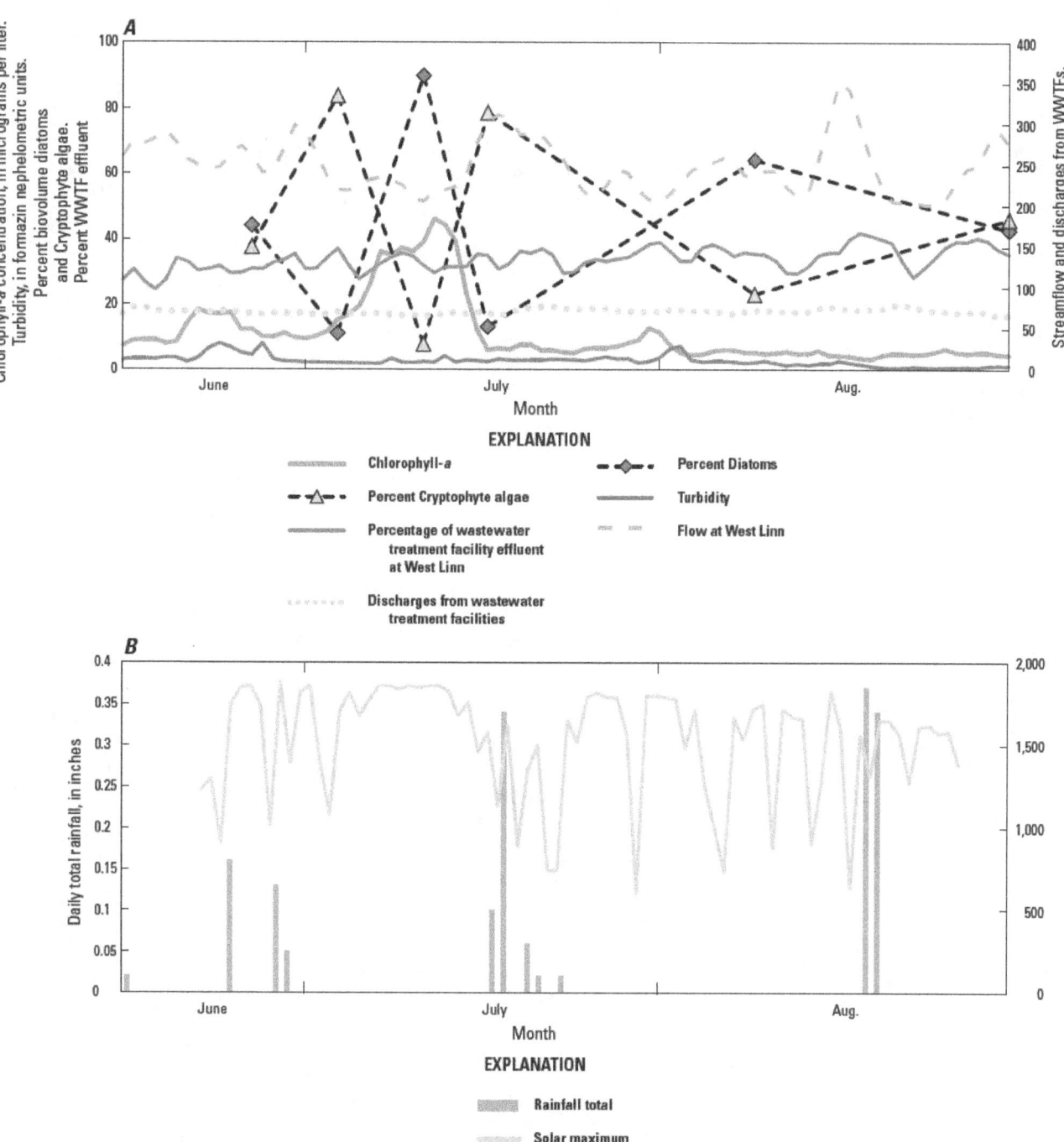

Figure 35. Time series of (*A*) chlorophyll-*a* and turbidity in the Tualatin River at the Oswego Dam (river mile [RM] 3.4), and percent biovolume of diatoms and cryptophyte algae at Stafford Road (RM 5.4), and (*B*) rainfall and solar radiation at the Durham wastewater treatment facility (WWTF), Oregon, June–August 2007. Daily average chlorophyll-*a* and turbidity from the Oswego Dam monitor (RM 3.4).

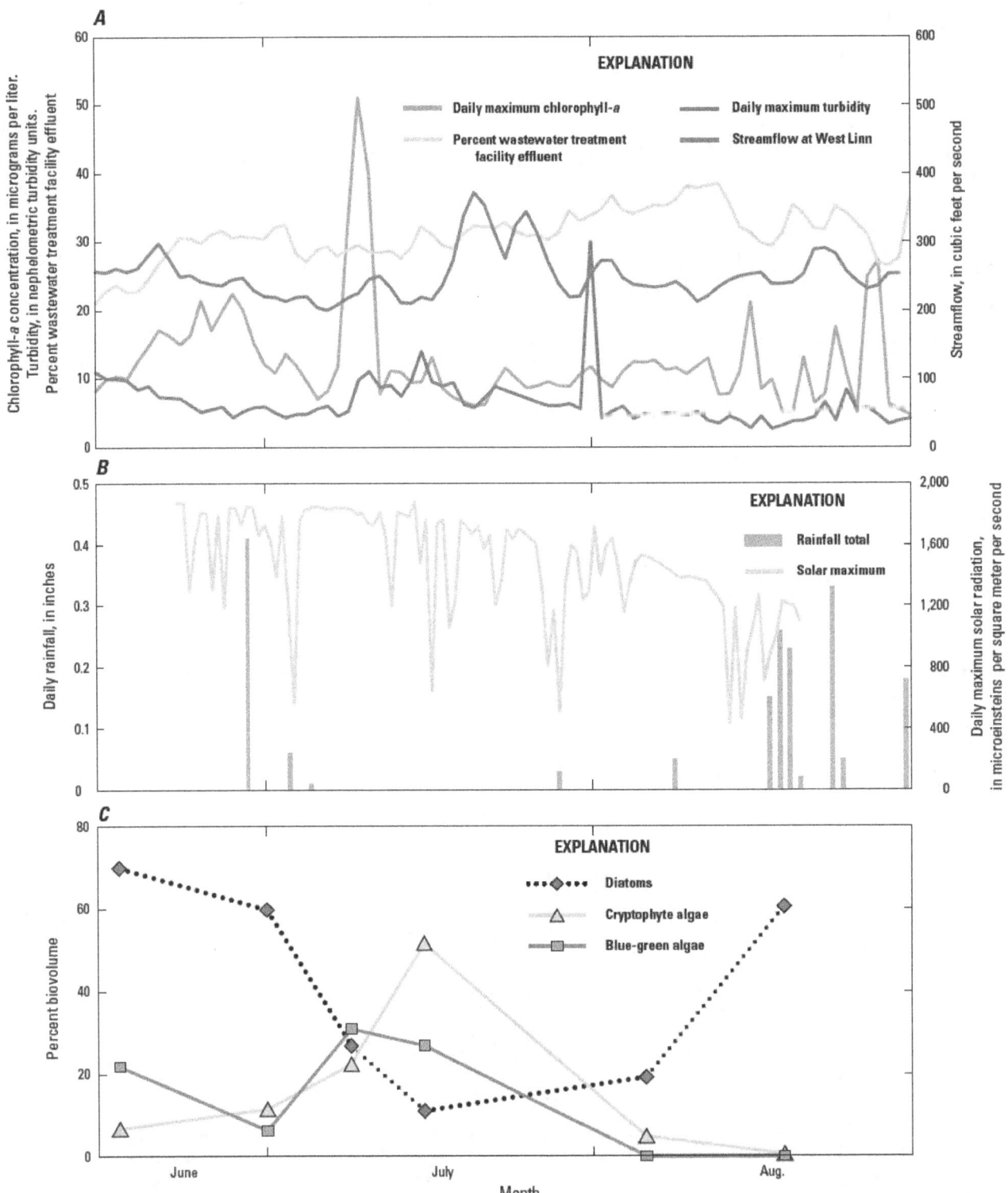

Figure 36.   Time series of (A) chlorophyll-a and turbidity in the Tualatin River at Oswego Dam (river mile [RM] 3.4), streamflow at West Linn (RM 1.8), and percentage of wastewater treatment facility (WWTF) effluent; (B) rain and solar radiation at the Durham WWTF; and (C) biovolume of diatoms, cryptophyte algae, and blue-green algae at RM 24.5 near Scholls, Oregon, June–August 2008.

In the first week of July, Chl-*a* concentrations in the upper river at Highway 219 (RM 44.4), upstream of Rood Bridge, sharply increased from 2.1 to 6.7 µg/L, and by then concentrations had also sharply increased—to 65 µg/L at Elsner downstream. By the following week, Chl-*a* concentrations had increased to 24 µg/L at Highway 219 and 83 µg/L at Elsner. By July 28, the Chl-*a* concentration at Highway 219 was 46 µg/L, but after that date it declined over a period of about 6 weeks. The bloom dissipated once pumping from Wapato ended and the inocula associated with those discharges stopped entering the river. After the bloom ended, the daily minimum DO concentration at the Oswego Dam declined to less than 6.5 mg/L for 60 of the next 62 straight days, falling to 3.8 mg/L on August 20. By the time diatoms made a bit of a comeback in late August, more rain and clouds ended the blooms. Given the severity of the July blue-green algae bloom, the low DO values in August probably resulted from a combined reduction in algal photosynthesis and DO demand from bacterial decomposition of senescing algal cells overlaid on the background SOD. Not since 2003 had the DO concentration been that low (table 14).

Evaluation of Hypotheses to Explain Phytoplankton Declines

Based on the evaluation of the bloom–crash sequences for 2003–08 and the multivariate analyses of the 2006–08 data, six factors are hypothesized to cause bloom crashes or prevent blooms from rebounding in August. These include: (1) light limitation, (2) a reduction in the inocula, or amount of phytoplankton entering the lower river from upstream sources, (3) increased summer streamflows, (4) changes in the dominant sources of flow, as an increasing percentage of flow augmentation and WWTF effluents has decreased the percentage of natural flow, (5) zooplankton grazing, and (6) low concentrations of bioavailable phosphorus. Each of these six proposed hypotheses is evaluated below.

Light Limitation

Hypothesis 1: Reductions in phytoplankton biomass and bloom crashes are brought about by light limitation caused by clouds or high turbidity.

Light is a required factor for algal photosynthesis, and light may become a limiting factor for phytoplankton growth during cloudy periods or when high turbidity limits light penetration into the water column. Rain and cloudy weather was a possible contributing factor in 6 of the 10 blooms evaluated for 2003–08 (table 15). Support for this hypothesis comes from numerous observations of algal declines during inclement weather, especially since monitoring has included continuous data for DO, pH, Chl-*a*, solar radiation, and rainfall that greatly facilitates the tracking of blooms and deciphering the causes of declines or crashes (fig. 35). Light limitation has been known as an important factor affecting algal blooms in the Tualatin River since the early 1990s and has been verified as an important factor not only through the available data but also through modeling analyses (Rounds and others, 1999; 2001). Often during spring and early summer, even partly cloudy days can have a marked effect on algal photosynthesis and DO production, but once sunny days return, the algae typically rebound, Chl-*a* levels increase, and DO production resumes.

Light limitation is also suggested at some level from the pattern in Chl-*a* and turbidity observed at Elsner (RM 16.2; fig. 27), which shows an increase in the algal biomass when turbidity is less than about 10 NTUs, with the bulk of the highest values occurring at even lower turbidity levels. This may help to explain potential decreasing trends in Chl-*a* concentrations in the upper part of the Tualatin River basin (fig. 7); more research and data are needed to clarify turbidity trends in these upstream areas. Elevated turbidity has been known to limit algal activity in the Tualatin River, as the data from 1996 at Elsner show (fig. 25); in that year, algal growth was suppressed to such an extent that maximum pH levels in the lower river were far lower than in the several years preceding or subsequent.

The apparent influence of turbidity on algal Chl-*a* means that if some small population of algae is needed to help maintain minimum DO levels, then erosion control becomes more important in this silt-laden watershed, especially during spring and summer when algal populations are developing. Unfortunately, this is also the period when some wetlands are drained and many other processes that may generate turbidity occur. Higher turbidity in the Tualatin River occurs from rain storms and urban runoff mobilizing materials from the tributaries, from drainage of wetlands for agricultural use, from higher flows associated with flow augmentation and reservoir releases, and from other sources. Even small reductions in algal populations in the upper part of the basin may lead to large algal declines in the lower river, which can result in low DO concentrations. In summary, this hypothesis seems to have some merit and basis in fact, and future research and monitoring of issues related to this topic would be valuable.

Reduced Algal "Seed Source" to Inoculate the River

Hypothesis 2: Reductions in upstream algal inocula "seed sources" help to account for the decline in algal populations.

For phytoplankton populations to develop in the river, an initial seed population of viable algal cells must be present and delivered to the river. This occurs either by germination of algal cells from river bottom sediments (Stoermer and Julius, 2003), or by immigration of an inoculum from upland sources such Barney Reservoir, Hagg Lake, small ponds, tributaries, wetlands, drained agricultural areas (Wapato Lake, for example), and/or other instream or off-channel habitats that support algal growth. These inputs may enter the mainstem through tributary discharges, by pumpage, or by resuspension. Centric diatoms, for example, can live among bed sediments in backwater habitats or on submerged wood, where they develop long filaments. Such filaments fragment easily, become entrained into the water column when disturbed, and may be an important source of algal inocula for the river.

A rich diversity of algae was available during summer to seed the lower river as far upstream as RM 24.5, where the highest algal species richness for any site (91 taxa) occurred during the 3-year study. In comparison, downstream sites had lower algal taxa richness, ranging from 58 to 70 taxa. Many of the taxa found only at RM 24.5, however, were benthic diatoms such as *Nitzschia*, *Navicula*, *Caleonis*, *Pinnularia*, *Amphora*, and others, which would not be expected to proliferate in the lower-gradient reservoir reach downstream. The next upstream sample site (Rood Bridge, at RM 38.4) had fewer taxa (67) but species similar to the sites downstream from RM 24.5. This increase in algal taxa richness between RM 38.4 and RM 24.5 could be attributed to inputs from any number of tributaries, including Rock Creek (RM 38.1), Gordon Creek (RM 37.4), Butternut Creek (RM 35.7), Christensen Creek (RM 31.9), Burris Creek (RM 31.6), Jackson Creek (RM 30.8), and Baker and McFee Creeks (RM 28.2). Many of these creeks drain areas with ponds that could serve as important sources of inocula to the lower Tualatin River; an examination of those sources might be a fruitful topic for future study.

The degree to which algal populations develop in the upper river has a direct effect on the eventual size of the population downstream. Large inputs of algal seed sources can provide the initial materials that produce large blooms downstream, depending on the location, flow rate, and other factors, as the 2008 Wapato Lake event demonstrated. In that case, elevated concentrations of algae, zooplankton, SRP, and ammonia initiated and fueled a large bloom of blue-green algae in the lower river; this was the largest bloom in recent years. This situation is not unique: algal blooms in the Sacramento River, California, are enhanced partly by the injection of algal inocula and nutrient rich water from upstream agricultural drains (Stringfellow and others, 2009).

Disruption of the seed source, if it occurs, would make it unlikely for algal populations in downstream reaches to rebound until another seed population enters the system from upstream, which could take time. A good example of this occurred at the end of the Wapato Lake event, when a reduction in the amount of algae coming into the upper river was the beginning of the end of the bloom (Bonn, 2008). Declines in algal populations in the lower Tualatin River during 2006–07 were preceded by declines in Chl-*a* upstream at Rood Bridge, where concentrations dropped from 3–5 μg/L to 1–2 μg/L prior to the decline in algal biomass downstream.

In all, it took about 6 weeks for the 2008 Wapato bloom to dissipate once the upstream source was cut off. Future targeted sampling of algae in the water column and bed sediments in the upper basin at key times and places could help to identify how the river is seeded with algae so that a better understanding of the dominant processes can be identified.

Higher Summer Streamflows

Hypothesis 3: Higher summer streamflows have reduced algal growth by lowering residence time and possibly also lowering water temperatures.

It is well established in the Tualatin River (and for many other rivers worldwide) that streamflow has a large effect on phytoplankton, primarily by affecting the amount of time available for algal populations to develop, grow, and reproduce. This effect is evident in the Tualatin River (fig. 37). In general, flows less than about 300 ft³/s at West Linn (RM 1.8) result in a long enough residence time (7–10 days) through the pooled reservoir reach for algae to multiply into substantial populations, provided that growing conditions are favorable. Streamflow in the Tualatin River during summer can be as low as 100–200 ft³/s, which provides as many as 14–17 days for phytoplankton to develop into a large bloom.

Withdrawals through the Oswego Canal at RM 6.7 decreased streamflow by about 50 ft³/s during summer prior to 1996, resulting in substantially increased residence times in the several miles upstream of Oswego Dam; in recent years, such withdrawals are largely inconsequential, resulting in higher flows and less time for algal growth. In addition, the use of flashboards at the Oswego Dam, which was commonplace in the 1980s and early 1990s but not in recent years, raised the water level approximately 10 in. in the 6-mi reach upstream of the dam, which further increased the residence time. Because this practice was discontinued, residence time has been further reduced in recent years.

If flows are high enough, algal populations may be "washed out" (Reynolds, 1990). This was demonstrated in September 1993, when a release from Hagg Lake, conducted as part of an experiment, increased flows and greatly lowered Chl-*a* concentrations (fig. 38). Increased flows in July and especially August (figs. 29 and 30) in recent years have reduced travel time such that algae now have less time to grow. The discontinued use of flashboards in the last 10 years also has the effect of reducing the residence time, but it is probably not as big a factor as the increased flow.

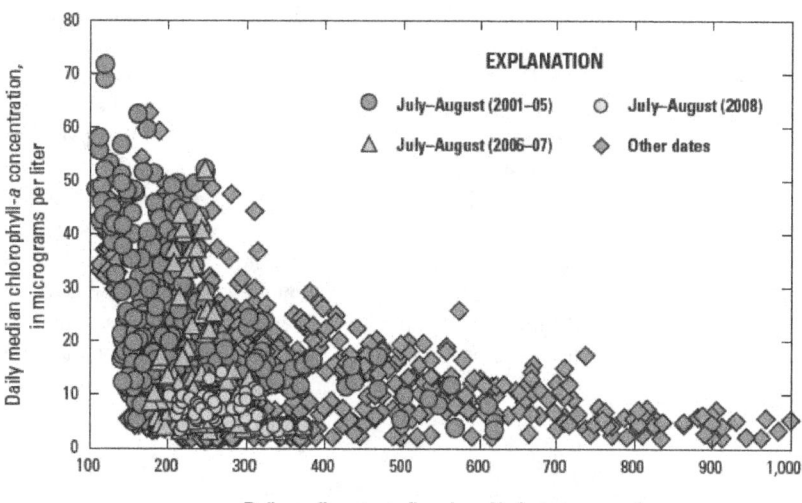

Figure 37.   Streamflow and chlorophyll-*a* in the lower Tualatin River at the Oswego Dam (river mile 3.4), Oregon, 2001–08.

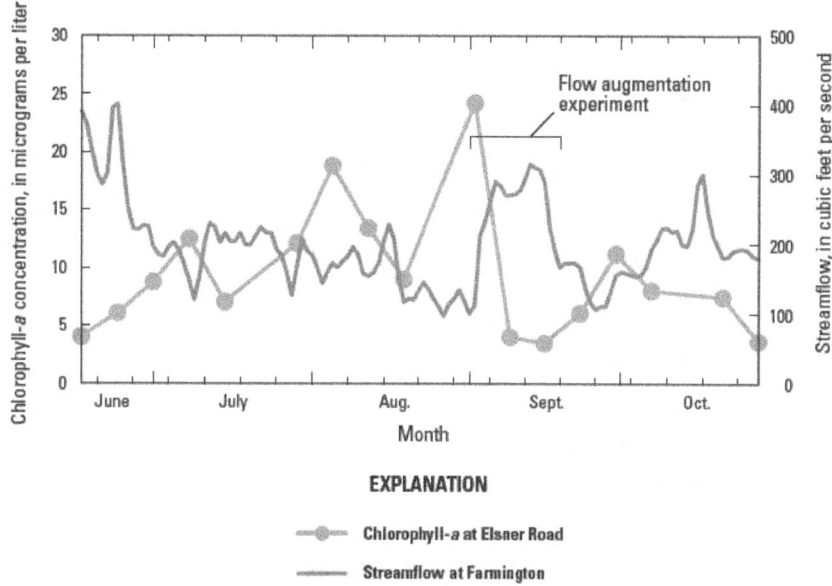

Figure 38.   Effect of the September 1993 Hagg Lake flow augmentation experiment on chlorophyll-*a* concentrations in the Tualatin River at Elsner Road (river mile [RM] 16.2), Oregon. Streamflow data from Farmington (RM 33.3).

Changes in the Dominant Sources of Flow

Hypothesis 4: Changes in the dominant source of flow during July and August may contribute to declines in algal populations.

Changes in the dominant source of flow in the Tualatin River, which include greater contributions from reservoir flow augmentation and discharges from the WWTFs, especially the Rock Creek facility, might also partly explain the reduction in algal growth in recent years. In 2006–08 the percentage of natural flow in the river declined from about 70–80 percent of the total flow in June to 38–40 percent later in summer (fig. 12) in response to reservoir releases. This reduction in

natural flow is anticipated to reduce algal populations by dilution and, thereby, decrease the concentration of algal inoculum reaching the lower river (see hypothesis 2). Given that both Chl-*a* and diatom biovolume tend to increase with higher percentages of natural flow at the two uppermost sites sampled, RM 38.4 and 24.5 (fig. 31), a change toward less natural flow might be partly responsible for algal declines. There is, however, no clear trend in the percentage of natural flow over the past twenty years (fig. 39), which is not entirely unexpected as natural flows are governed by multiannual patterns in precipitation, runoff, and groundwater discharges, as well as year-to-year variations in flow-augmentation management.

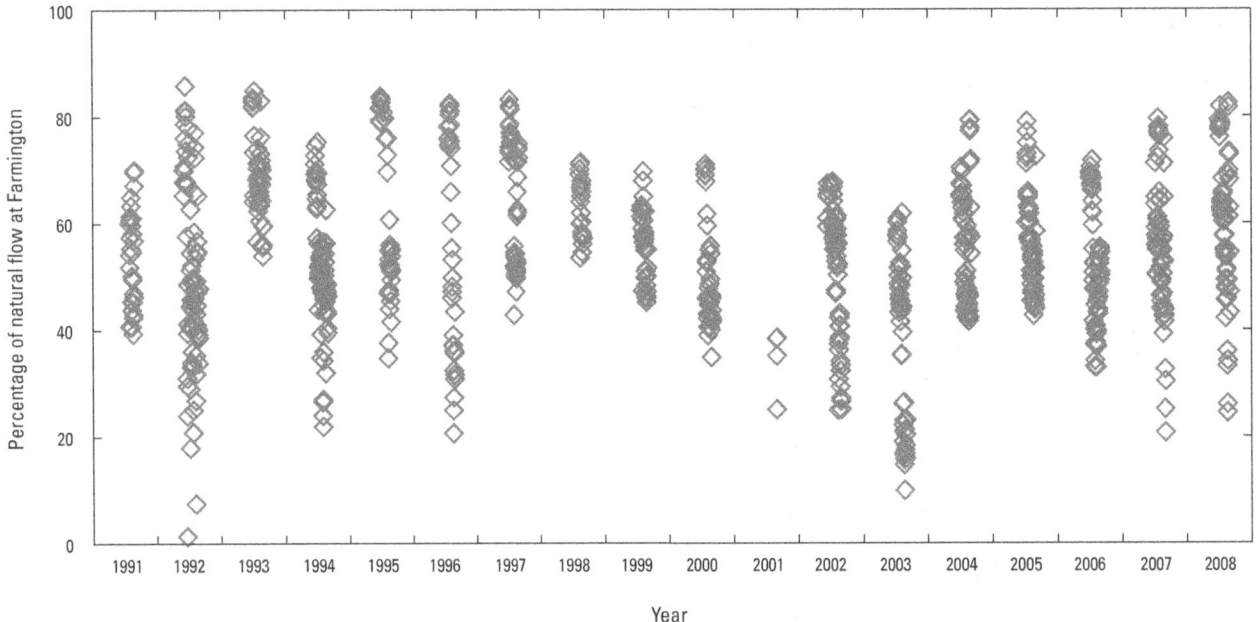

Figure 39. Percentage of natural flow in the Tualatin River at Farmington (river mile 33.3), Oregon, June–August 1991–2008.

The importance of flow was clearly evident in the multivariate BEST models (table 11), which identified total flow and the source of flow (including flow augmentation, natural flow, and WWTF effluent as a percentage of total flow) as the most significant factors that define patterns in the phytoplankton assemblages. That the dominant source of flow was important makes sense given the potentially large differences among these sources in terms of the amount and type of algae they would be expected to contribute, and the differences in the quality of each source.

It is not known, but certainly possible that the amount of algae in the natural component of the flow coming from upstream also has decreased along with the amount of natural flow. Even small reductions in Chl-*a* levels far upstream could give rise to lower populations downstream, especially if travel times are shortened by higher flows during summer, which has occurred in recent years, from periodic pulses of flow augmentation water or WWTF discharges (figs. 29, 30).

Differences in the flow source had notable effects on the relative biovolume of just a few key taxa, namely *Stephanodiscus binderanus* and *S. hantzschii*, which show increased prevalence at higher natural flows and decreased prevalence at higher levels of flow augmentation, whereas *Cryptomonas erosa* and *Chlamydomonas* sp. were higher during augmentation (fig. 40). Note, however, that although the relative biovolume of these two *Stephanodiscus* taxa increased with higher amounts of natural flow, the overall biovolume of algae was reduced with higher flow. The highest biovolume for these two taxa occurred when flows were between 200 and 300 ft³/s. Note also that the levels of natural flow and flow augmentation are highly correlated with the time of year, length of day, and other factors, so these results may not necessarily represent true cause and effect, and additional research and analysis is needed.

Water Storage Reservoirs and Flow Augmentation

Water releases from Hagg Lake (primarily) and Barney Reservoir (fig. 1) have a marked effect on the river's flow and quality. By increasing the rate of flow, the travel time through the reservoir reach (RM 33–3.4) decreases, reducing the amount of time algae have to grow.

The decline in Chl-*a* during the 1993 flow augmentation experiment, and also at the upper sampling sites (Rood Bridge [RM 38.4] or RM 24.5) in late July during this study, suggests that upstream reservoir releases probably are not a significant source of inocula in mid- to late-summer. Releases from Hagg Lake in summer are from an outlet located at a depth of about 70 ft at full pool (Sullivan and Rounds, 2005), well below the thermocline from late spring through mid- to late-summer, which is an important factor that might limit the export of plankton from the lake. Although it is possible that the

reservoirs help seed the Tualatin River with algae during the growing season—about half of the algal taxa found in the river also have been identified in Hagg Lake—this possibility has not been investigated. Alternatively, releases from upstream reservoirs could dilute the algal inocula from other sources to such an extent that downstream populations do not thrive, as the negative correlation between flow augmentation and Chl-*a* suggests (fig. 31).

The effect of flow augmentation by itself was highly variable, although algal biovolume was notably lower when flow augmentation was greater than about 95–100 ft³/s. The biovolume of diatoms at the two uppermost sites, for example, was negatively correlated with flow augmentation (fig. 31), whereas higher relative biovolumes of *Cryptomonas erosa* and *Chlamydomonas* sp. occurred with greater amounts of flow augmentation (fig. 40). This was not the case in 1993, when the abundance of *Tabellaria fenestrata* (a colonial diatom) and *Melosira varians* increased during the experimental flow release, whereas *Cryptomonas erosa* abundance remained unchanged. Although it is likely that algal assemblages in the lower Tualatin River are affected by upstream seed sources including releases of phytoplankton from Hagg Lake, additional data are needed to quantify the seasonal inputs of algae from all potential sources before this hypothesis can be sufficiently evaluated.

WWTF Discharges

The proportion of treated wastewater in the Tualatin River has increased over the past couple decades (fig. 13), and despite its high quality, that increase may be contributing to algal declines. The BEST analysis identified the percentage of WWTF effluent to be important in determining patterns in algal assemblage, although percent WWTF effluent was not as important as total flow, and water temperature also was included in the solution (table 11). In nearly all cases, the total algal biovolume measured downstream from the Durham WWTF outfall location was lower than that measured just upstream of these inputs (fig. 41), particularly when biomass levels were high. This difference can be partly attributed to dilution from WWTF effluent discharges and small inputs from Fanno Creek, but may also be due to some particular quality of the treated effluent.

At times when flow from the WWTFs is greatest as a percentage of total river flow (as much as 38 percent in 2006–08), there was an accompanying increase in the relative biovolume of blue-green algae and a decrease in the relative biovolume of two *Stephanodiscus* species (fig. 42). Because the percentage of WWTF effluent is always highest in late summer when flows are low, it is impossible to determine with the available data whether these observations were due to WWTF effluent, from natural seasonal succession in the plankton, or changes in upstream inocula, zooplankton grazing, reservoir releases, temperature, or some other factor.

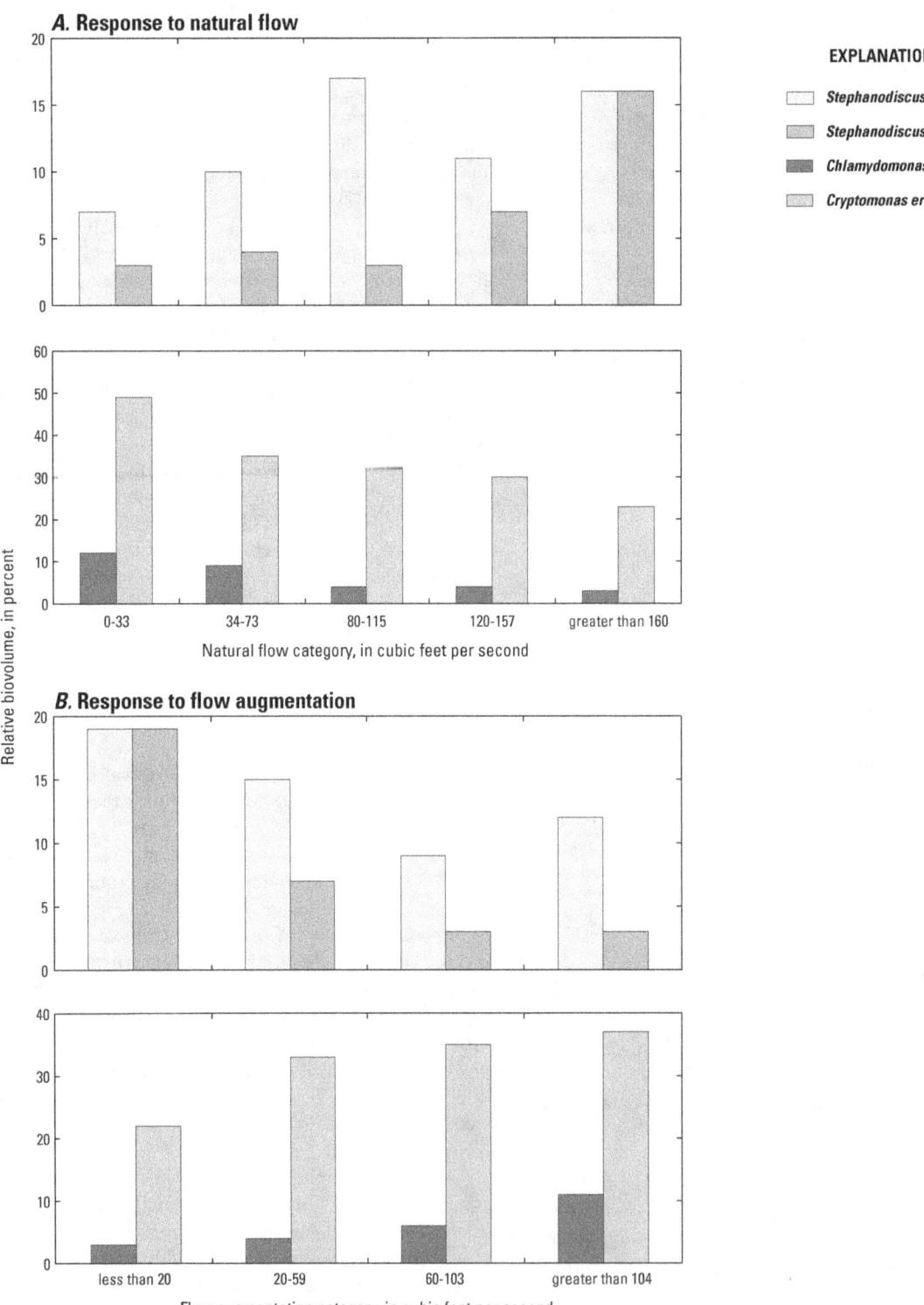

Figure 40.   Relative biovolume of select algal taxa showing potential (*A*) responses to natural flow and (*B*) flow augmentation in the Tualatin River, Oregon, 2006–08. Natural flow is not derived from reservoirs or WWTFs. Flow augmentation includes reservoir releases from Hagg Lake and Barney Reservoir minus estimated withdrawals.

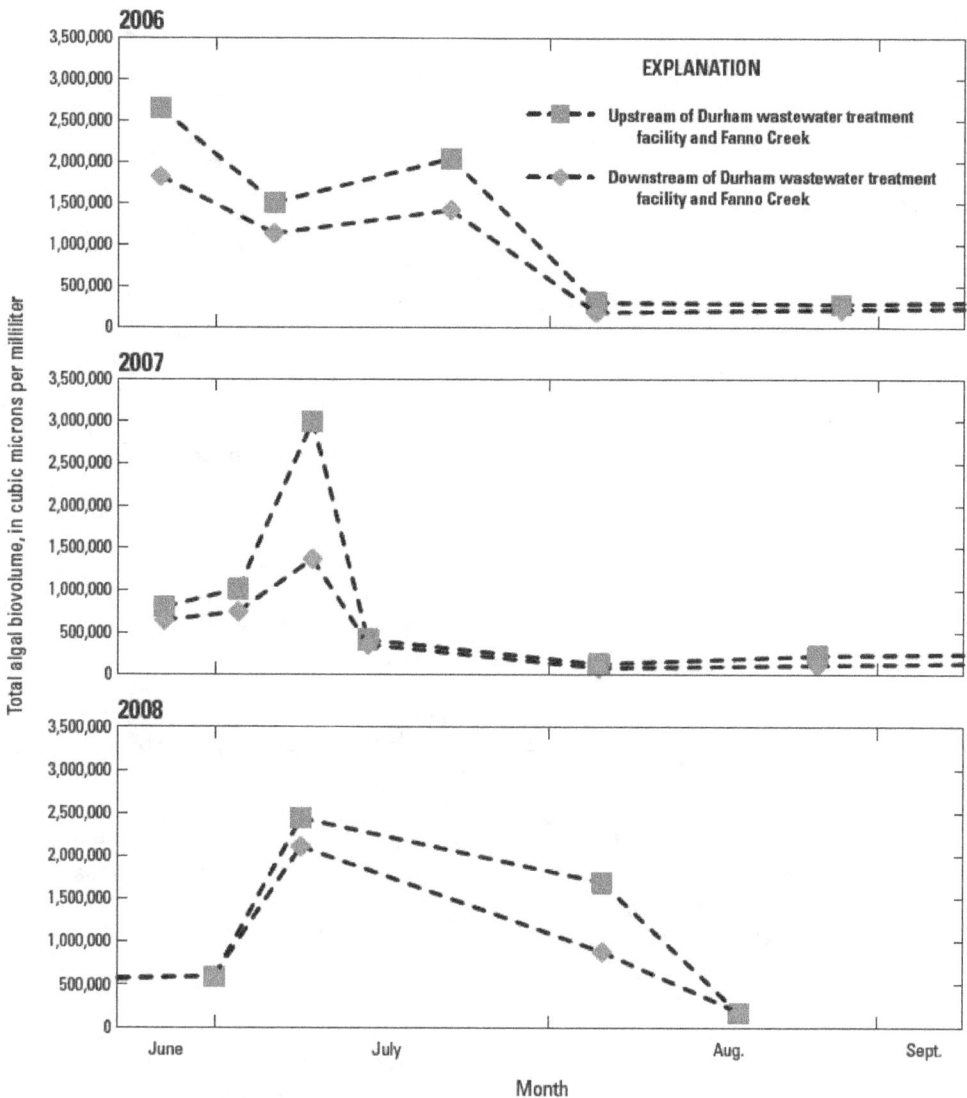

Figure 41. Comparison of total algal biovolume in the Tualatin River upstream and downstream from the Durham WWTF and Fanno Creek, Oregon, 2006–08.

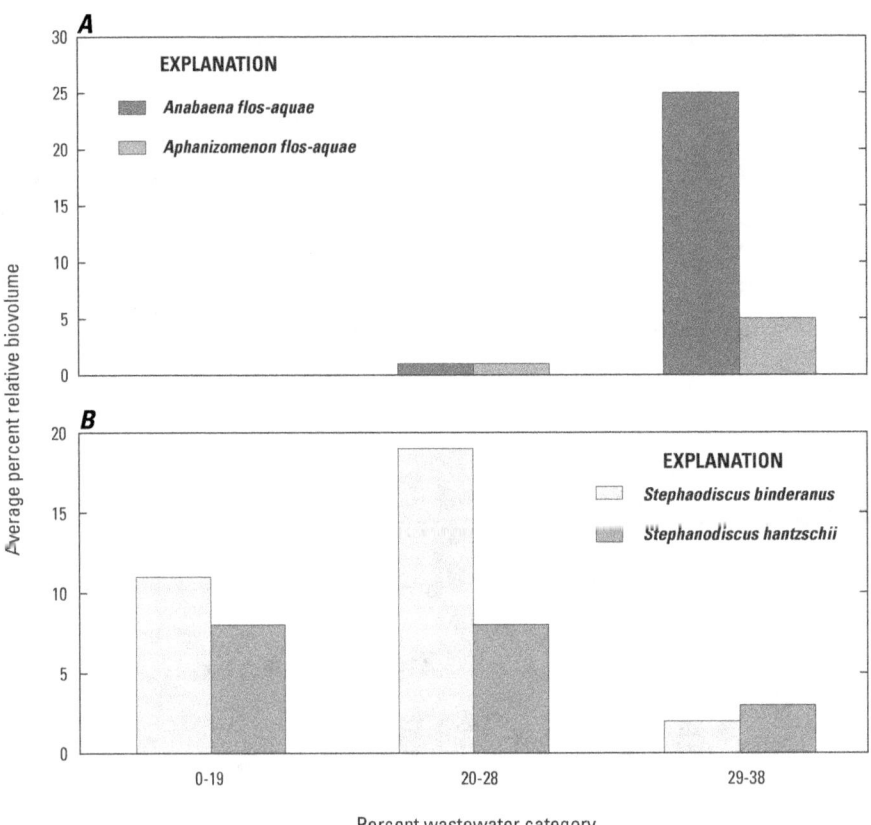

Figure 42. Average percent relative biovolume of select taxa showing potential (*A*) positive responses and (*B*) negative responses to percentage of wastewater treatment facility (WWTF) effluent in the Tualatin River, Oregon, 2006–08.

The limited bioassay experiments conducted for this study were inconclusive; while some stimulation of phytoplankton growth was observed at 30 percent WWTF effluent concentration, effluent concentrations of 50 percent sometimes increased and other times decreased Chl-*a* and DO production during the experiments. Because most of the bioassays were conducted during the unusual Wapato Lake discharge event, results from only three experiments were presented here—with no or limited replication in treatments. Data from these initial pilot experiments should not, therefore, be considered in any way definitive.

Overall, the hypothesis of a connection between algal declines and the source of flow can neither be accepted nor rejected at this point. The available data certainly point to some interesting correlations of algae with the source of flow, but definitive cause and effect has not yet been confirmed. The possibility that WWTF effluents could be decreasing algal populations remains unanswered, and the simple dilution of the natural flow could certainly contribute to the observed declines in Chl-*a*, but other causes are also likely. The inocula hypothesis (2), for example, is supported by multiple lines of evidence, and the source of flow (this hypothesis) may act to reinforce the processes involved in the inocula hypothesis.

Zooplankton Grazing

Hypothesis 5: Zooplankton grazing accounts for some algal declines.

Although not measured directly in this study, zooplankton grazing can have a strong effect on phytoplankton abundance and species composition in lakes and rivers (Wetzel, 1983; Thorp and Covich, 2001). Previous modeling of Chl-*a* in the Tualatin River (Rounds and others, 1999) found that, at times, the simulated biomass was higher than that measured in the river, and attributed this to losses from zooplankton grazing. A comparison of zooplankton taxa from 2006–08 with previous data from 1991–93 (Doyle and Caldwell, 1996) showed that many of the same Genera (*Daphnia*, *Ceriodaphnia*, and *Bosmina*) were dominant in all years.

In spring or summer, before zooplankton populations are significant, phytoplankton growth occurs without much control by zooplankton. Zooplankton abundance increases as flows decline and food resources become more abundant. As zooplankton densities increase, grazing losses can begin to surpass the reproduction rate of the phytoplankton, causing algal populations to decline. With fewer food particles available, sharp reductions in the zooplankton density occur, tracking or lagging declines in the phytoplankton (fig. 19). By August, with fewer phytoplankton and zooplankton in the water column, a clear-water phase can occur, similar to that exhibited in many temperate lakes during late summer (Wetzel, 1983). In the Tualatin River at Stafford Road, for example, water transparency (Secchi disk depths) increased by about 2 ft during these late-season periods—especially in 2006 and 2007, years when zooplankton grazing was suggested by data or observations of high amounts of zooplankton in net tows.

Supporting evidence for this hypothesis comes from several observations, including high overall density of zooplankton in the lower river at certain times (fig. 17). In 2006 and 2007, zooplankton abundance increased downstream of RM 24.5 with notably higher densities at and downstream of the Highway 99W/Jurgens Park sites near RM 11. In 2008, as a result of discharges from Wapato Lake, large densities of copepods developed in the river, resulting in much higher densities at RM 24.5 (fig. 17).

More convincing evidence is found in the relationship between zooplankton density and total algal biovolume (fig. 19), which suggests zooplankton grazing may, at times, contribute to algal declines. Without specific measurements of grazing rates, however, it is difficult to infer how grazing affects phytoplankton abundance or species composition. Many of the zooplankton taxa found in the Tualatin River have been shown in other studies to graze on many of the phytoplankton taxa that occur in the Tualatin River. For example, laboratory experiments by Fulton (1988) showed that filamentous centric diatoms (*Melosira granulate*) were readily consumed by cladocerans including *Bosmina longirostris*, which was one of the dominant zooplankton species in the Tualatin River in 2006–08, occurring in 94 percent of samples (table 7). Descy (1993) documented a sharp 10-fold decrease in phytoplankton abundance in the River Moselle (France) that was attributed to zooplankton grazing. Some of the most heavily grazed diatoms in that study (*Stephanodiscus* and *Cyclotella*) were often abundant in the Tualatin River.

Algal types vary in their susceptibility to zooplankton grazing, and many have adaptations that make them less palatable. For example, many of the diatoms in the Tualatin River form colonies, spines, or other structures that may deter grazing. Diatoms are generally a preferred food source over most blue-green algae, which are commonly regarded as a poor-quality food source for zooplankton (Gulati and DeMott, 1997). While diatoms are rich in fatty acids, blue-greens are considered to be less desirable to grazers because of their larger size, presence of a mucilaginous coating, and potential for toxin production (Caramujo and others, 2008). Similar to diatoms, small naked flagellates, including *Cryptomonas erosa*, also are considered to be a higher-quality food compared to blue-green algae (Chen and Folt, 1993).

While zooplankton grazing certainly exerts a negative effect on algal populations, the strength of the grazing hypothesis needs to be tested with experimental approaches that elucidate and quantify the grazer-algal relations in the Tualatin River. Ideally, such experiments would also include planktivorous fish, as they may have an important influence on the abundance or composition of zooplankton.

Phosphorus Limitation

Hypothesis 6: Phytoplankton bloom declines are caused by limited availability of bioavailable phosphorus.

Phosphorus limitation may be a contributor to algal declines and the crashes of certain algal blooms, as concentrations of SRP often were reduced to very low levels, less than 0.01–0.15 mg/L, during high biomass peaks. As just one example, SRP concentrations were decreased to 0.01 mg/L at Boones Ferry (RM 8.7) in July 2004 during an algal bloom (fig. 43); similarly low concentrations of SRP occurred in the lower river during 6 of the 10 bloom crashes listed in table 15. In addition, the near linear relation between Chl-*a* and SRP demonstrates this *negative* or "uptake" pattern (fig. 15*B*).

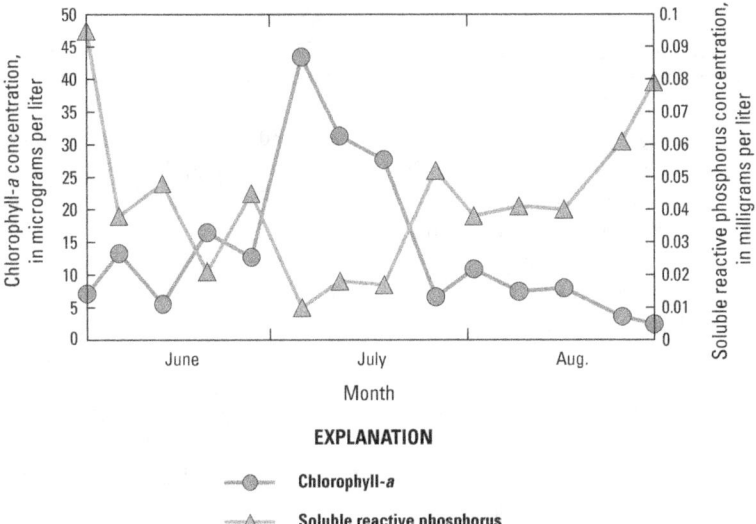

Figure 43. Pattern in chlorophyll-*a* and soluble reactive phosphorus concentrations in the Tualatin River at Boones Ferry (river mile 8.7), Oregon, June–August 2004.

Previous studies and modeling (Rounds and others, 1999) found that phosphorus concentrations essentially place a cap on the size of the blooms, which is consistent with these results, but they also found that typical SRP levels were sufficient to initialize a bloom, and so low SRP does not always explain why algal growth remains low. The importance of SRP to phytoplankton populations was clear from the BEST analyses, which identified SRP as an important factor in several of the solutions, and the most important factor (highest rho value, 0.242) in the solution for the combined 2006–08 years that was significant at P<0.05 (table 11).

Although phosphorus concentrations typically may be sufficient to start a bloom, algal uptake can decrease SRP concentrations to levels that may become limiting. Despite the fact that many of the algal taxa in the Tualatin River are considered eutrophic (table 6), lab experiments (VanDonk and Kilham, 1990) showed that some of the same diatom species in the Tualatin River (*Stephanodiscus hantzschii*, *Asterionella formosa*, and *Fragilaria crotonensis*) are able to assimilate SRP at very low concentrations and have relatively low half-saturation growth constants for phosphorus, ranging from 0.001 to 0.003 mg/L, which makes these taxa highly competitive during periods of low SRP availability. The presence of these taxa could indicate that the phosphorus supply, while perhaps limiting to other algae, may not limit the growth of these algal species if the measured SRP concentrations are an accurate measure of bioavailable P.

Measured summer SRP concentrations in the Tualatin River were greatly decreased as a result of major upgrades at the WWTFs in 1990–1992 and subsequent improvements in treatment processes. Median SRP concentrations in treated effluent for the May–October period in 2006–08 ranged from 0.013 to 0.018 mg/L (table 4), much lower than typical concentrations occurring in the Tualatin River. As a result, SRP concentrations can be quite low in the river in late summer when WWTF effluent comprises a higher proportion of the flow or during phytoplankton blooms when uptake rates are high.

Whether the SRP in the river is bioavailable, however, continues to be an important question. Results from Tualatin River studies in the early 1990s showed that phosphorus could form a co-precipitate with colloidal iron and silica, but the solubility and bioavailability of that coprecipitate is unknown (Mayer and Jarrell, 1995). Future studies of the forms and bioavailability of phosphorus in the Tualatin River and in treated effluent would provide useful information. For example, Li and Brett (2010) found that much of the SRP in the Spokane River, Washington, another river that receives substantial amounts of municipal wastewater, was not readily available to at least some types of algae tested in bioassays. While phosphorus reductions are an important management strategy to reduce the occurrence, frequency, and magnitude of algal blooms, and potentially toxigenic harmful algal blooms, such aggressive treatment may be starving potentially beneficial algal populations and affecting them or their environment in ways that are not clearly understood.

Conclusions and Implications for River Management

The results of this study indicate that algal populations in the Tualatin River are influenced by many factors that are important at various times and locations. These factors include the magnitude and sources of streamflow, available light, instream turbidity, upstream inocula, grazing by zooplankton, and the amount and bioavailability of phosphorus. Although substantive algal blooms occur nearly every year, these factors help to determine when blooms form, how large they become, how long they last, when and how the bloom declines or ceases altogether, and which algae species are dominant. Many of these factors have changed since 1991 because of imposed regulations, population growth, expansion of an upstream reservoir, and alterations in flow management, to name just a few influences. The combined effect of these factors on algae generally has been to decrease algal populations, but particularly in late July to September and to such an extent that the loss of photosynthetic activity has led to problematic DO concentrations, especially since 2003.

Several possible causes were identified to explain the decreased phytoplankton abundance in the lower river in recent years. First, the level of algae in the river, estimated from phytoplankton chlorophyll-a (Chl-a) concentrations, has declined in the upper river. While such a decline alone would almost certainly reduce the size of downstream populations, the simultaneous dilution from increased reservoir releases for flow augmentation and higher discharges from the Rock Creek wastewater treatment facility (WWTF) also work to reduce algal populations. In addition, a slightly higher turbidity in the upper river may be important in causing an increased occurrence of light limitation. Soluble reactive phosphorus (SRP) concentrations, the percentage of streamflow from various sources (natural flow, flow augmentation, and WWTF effluent), and zooplankton grazing all were identified by the multivariate BEST analyses as important factors for explaining variations in the phytoplankton species composition; all these were listed as possible factors contributing to bloom crashes in 2006–08.

It is clear from data and modeling that streamflow exerts an important influence on phytoplankton abundance in the Tualatin River by affecting residence time, but the source of that flow also is important. The fact that flow variables were the most important factors in the top two multivariate BEST models suggests that it is the quality of flow, not simply the magnitude of flow (and its effect on residence time), that shapes phytoplankton assemblages. This influence could be due to a change in the amount or quality of phytoplankton inoculum entering the lower reservoir reach, the turbidity or SRP level of the water, or some other property. All such characteristics could have implications for downstream algal populations, photosynthesis and dissolved oxygen (DO) production.

Algal growth limitations due to low phosphorus concentrations probably did not account for the general algal declines observed in the Tualatin River since 2003 unless phosphorus bioavailability also has changed. As previous work (Rounds and others, 1999) has demonstrated, decreased SRP concentrations, at a minimum, can place a cap on the size of the blooms. Important questions remain, however, regarding the extent to which the SRP in the river and in WWTF effluent is bioavailable.

This study demonstrated the value of a long-term, basin-wide, routine monitoring program in helping to decipher longitudinal patterns in river conditions and trends over time. Such data will likely continue to advance an understanding of how the Tualatin River functions with respect to its plankton communities, and the data may help explain future changes or potential causes of water-quality problems. Only by considering data over the long-term can patterns in year-to-year variability be understood, and unusual years be identified—such as the 2001 drought or the 2008 "Wapato Event." Without such data, many more questions would remain regarding the annual cycles in phytoplankton populations in the river and reasons for the recent declines.

The results of this study can be used to refine future monitoring to confirm and further develop the hypotheses put forth here. Using that information, well-designed experiments and targeted sampling before, during, and after bloom crashes at key locations in the lower and upper reaches of the watershed should prove helpful in further understanding the sources, development, and dynamics of algal communities in the river. If phytoplankton photosynthesis is to be relied upon for maintaining minimum DO levels in the river, then a better understanding of algal sources and dynamics in the upper river is clearly needed.

If algal populations cannot provide sufficient DO to offset oxygen demands in midsummer under future flow conditions, then strategies aimed at reducing the sediment oxygen demand (SOD) might be the only other option—short of aerating the river or greatly increasing streamflow. A study using stable isotopes to characterize the sources of organic matter to bed sediments of the Tualatin River (Bonn and Rounds, 2010) determined that terrestrial plants and soils were the most important sources of organic matter to river sediments, implying a largely terrestrial source that sustains the SOD. Erosion control in a basin with large loads of fine sediment, however, can be difficult. New strategies might be needed to control erosion and stabilize streambanks. A study examining the sources and transport of organic matter in the Fanno Creek basin is using multiple techniques, including new fluorescence methods that may prove useful for identifying carbon sources in the watershed that might contribute disproportionately to the river SOD.

Building on results from this and other studies, a great deal of research still can be done on the factors affecting phytoplankton growth and population dynamics in the river, and the degree to which they are affected by flow quantity and sources, phosphorus availability, zooplankton grazing, temperature, and food web dynamics.

Possible Future Studies, Monitoring, and Research

While this study has identified several factors that appear to influence phytoplankton growth in the Tualatin River, there are additional areas of research that would deepen our understanding of how plankton assemblages in the Tualatin River function and respond. The following section provides some ideas and direction for future studies, monitoring, and research to fill data gaps, refine our understanding of plankton dynamics, and inform and enhance water-quality models that can be used by water managers to develop strategies that improve DO conditions in the lower Tualatin River.

Monitoring Needs

The current monitoring program implemented by Clean Water Services (CWS) and the U.S. Geological Survey (USGS) and other entities includes little monitoring of Chl-*a* or plankton communities upstream of Rood Bridge. Certain factors identified in this study, however, connect declines in phytoplankton in the lower river to conditions occurring in the upper river—such as the amount and characteristics of the algal population, the importance of various streamflow sources, and the amount and sources of turbidity. Additional biological sampling (Chl-*a* and algal identification and enumeration) at a few key sites in the upper basin would be useful for discerning the effects of the various flow sources in the basin and for monitoring the potential sources of algal inocula to the lower river. Additional sampling in the upper basin might also be necessary to identify the possible presence and causes of elevated turbidity that could be contributing to lower algal abundance in recent years. The current river monitoring program has, in the past, included algae and zooplankton sampling in the lower Tualatin River at Stafford Road (RM 5.5); that type of monitoring data should prove useful in characterizing plankton communities in that part of the reservoir reach. Other stations such as Elsner (RM 16.2) or Cook Park (RM 9.9) could be added to further enhance the monitoring program. These additions would generate data to help test or refine some of the hypotheses discussed in this report and strengthen our understanding of the factors that control phytoplankton populations in the upper and lower Tualatin River.

Periodic and targeted monitoring of plankton species and Chl-*a* (discrete or continuous) at sites in the upper basin such as Gaston or Cherry Grove, Wapato Creek, Scoggins Creek, and the Tualatin River between Dilley and Golf Course Road would provide critical information on the type and amount of algal inocula during the early, middle, and later parts of summer. This information could be used in an expanded multivariate data analysis or used to enhance water-quality models of the river in order to explore the effects of management strategies such as changes in the schedules of flow augmentation.

In addition to evaluating upstream reservoirs as sources of algal inocula, other sources including farm ponds, tributaries, and wetland drainage from areas including Wapato Lake, Fernhill Wetlands, and Jackson Bottom could be examined for their potential to contribute plankton to the Tualatin River. Periodic surveys conducted during spring and summer, for example, would be useful to document contributions of algae (and zooplankton) from these sources. The large effect of draining the Wapato Lake agricultural area in 2008 highlights the need to monitor this and other sources of upstream plankton inocula. Sampling of the Tualatin River at or near the Spring Hill Pump Plant, a site that is downstream of many important inputs such as Wapato Creek, Barney Reservoir, and Hagg Lake, could be used to initiate targeted sampling of these individual upstream sources when warranted.

Bioassay Experiments

Bioassay experiments, similar to the ones conducted in this study, could be used to answer questions regarding the effect of factors such as elevated turbidity, flow augmentation from various sources, and WWTF effluent on phytoplankton abundance. Such experiments can also be used to determine the level of phosphorus limitation, or to determine specific growth rates and other parameters used for input to water-quality models. While this study found that moderate levels of WWTF effluent may stimulate algal growth, higher concentrations (50 percent) may have either a reduced stimulatory effect or a negative effect. Additional bioassay tests with finer resolution of wastewater percentages could be used to verify and refine these interpretations.

Bioassays also could be used to determine whether SRP from various sources is available for algal growth. Although abundant data are available to determine the SRP levels in the river, it remains unclear whether that phosphorus is bioavailable (Mayer and Jarrell, 1995; Li and Brett, 2010). Detailed experiments that analyze the geochemical characteristics of the phosphorus, such as those used by Simon and others (2009), might be helpful. In addition, bioassay experiments using smaller increments of phosphorus to test samples collected during specifically targeted periods—such as when it appears that phosphorus might be limiting algal growth—might be useful for estimating phosphorus bioavailability.

Studies of Zooplankton—Phytoplankton Interactions and Fish Predation

Future studies of zooplankton–phytoplankton interactions, the effect of planktivorous fishes on zooplankton, and the cascade of effects that dictate the abundance and photosynthetic rates of the phytoplankton and their production of DO in the lower Tualatin River would be beneficial. Such experiments might also measure respiration rates of zooplankton and phytoplankton to gage their direct effect on DO. Future studies of zooplankton–phytoplankton interactions in the Tualatin River could provide a greater understanding of how plankton interact. It has been difficult to disentangle the effects of flow management and zooplankton grazing in some of the historical data; targeted studies of the zooplankton and their grazing effects would help to fill this gap in our understanding.

Little is known about the abundance of non-game fishes or their size/age distributions in the Tualatin River, but several fish species in the river are planktivorous when they are young, particularly nonnative species such as largemouth bass, bluegill sunfish, black crappie, pumpkinseed, yellow perch, and warmouth, which could have an important effect on zooplankton populations in the Tualatin River. Native planktivorous species include redside shiner, peamouth, and speckled dace. Previous USGS studies in the Tualatin River have found that small fish (less than 150 millimeters in length) were feeding on zooplankton. Future studies could examine the effect of planktivorous fish predation on zooplankton and the resulting effects on phytoplankton. A better understanding of these complexities could greatly improve existing or future water-quality models and river management strategies.

Artificial Neural Network Models

Although the artificial neural network (ANN) model developed by Rounds (2002) to predict DO concentrations at the Oswego Dam based on streamflow, sunlight, rainfall, and air temperature predicts DO with reasonable accuracy, the ANN model might be improved with additional inputs such as the proportions of natural flow, flow augmentation, and WWTF effluent, as well as ammonia loads from the WWTFs. Calibrating the ANN model with data on the size of the algal inocula entering the reservoir reach might also improve the accuracy of modeled phytoplankton population sizes. Other inputs to the model might include a seasonal component to the modeled zooplankton population, to better predict algal crashes and DO concentrations. Currently, the model performance is least successful during these periods.

ANN models are particularly well suited for problems in which large datasets—such as the available continuous water-quality, streamflow, and meteorological data (solar radiation and rainfall)—contain complicated nonlinear relations among many different inputs. Although mechanistic models that capture the essence of a system's instream processes often can be used to great advantage in evaluating management strategies, the complexities of the algal community in the Tualatin River and its ties to various flow sources and upstream inocula are somewhat limiting for such models. ANN models, on the other hand, can utilize a wealth of data and are not tied to any particular instream mechanism or factor; the key is in collecting the right kind of data and including those data in the ANN analysis. ANN models have been used successfully to predict the timing and magnitude of Chl-*a* in the Nakdong River (Korea) using similar types of data (Jeonga and others, 2001). An updated and more refined ANN model could be constructed for the Tualatin River to further evaluate and refine the hypotheses put forth here, or to develop new ones in future years.

Periodic Revisitation of the Multivariate Analyses

The multivariate statistical analyses of the combined biological, chemical, and streamflow datasets in this study proved useful in identifying several factors affecting the presence, abundance, seasonal patterns, and other characteristics of the algal community in the Tualatin River. As additional datasets are collected in the future, either through routine monitoring or targeted studies, it might be beneficial to revisit these multivariate analyses to determine whether the patterns and factors identified here are still important, identify any new influences, and detect trends or community shifts that might be tied to management strategies.

Acknowledgments

The authors wish to thank Jan Miller, Mark Poling, Robert Baumgartner, and Raj Kapur of Clean Water Services for their insights into the water quality of the Tualatin River and helpful comments on data interpretations. We are especially thankful to Jan Miller for her enthusiastic support for this project, and to Robert Baur for providing treated effluent for the bioassay experiments and insights into facility operations. We also appreciate the high-quality data produced by the Clean Water Services Water-Quality Laboratory and want to acknowledge and thank Steve Thompson and his staff for their dedication and willingness to analyze extra water samples from the USGS sampling trips. We thank Bernie Bonn for sharing her knowledge of the basin and providing data files on flows, reservoir releases and withdrawals used in the annual reports of the Tualatin River Flow Management Technical Committee. This analysis benefited from discussions with several people, including Peter Schmidt from the Tualatin

River National Wildlife Refuge and Tom Murtagh from the Oregon Department of Fish and Wildlife. We also want to recognize and thank several USGS staff, including Tammy Wood and Annett Sullivan, for sharing their knowledge of the basin, and Micelis Doyle, Michael Sarantou, Jami Goldman, and Kiara Smith (formerly with USGS) for helping with sample collection. Matt Johnston and Micelis Doyle are primarily responsible for operating the main-stem continuous water-quality monitors for the last 20 years, and their efforts to provide such a complete dataset is much appreciated. Lastly, we thank the staff of the USGS Portland Field Office and the Oregon Water Resources Department for maintaining the continuous streamflow gages in the Tualatin River basin.

References Cited

American Public Health Association, 1992, Standard methods for examination water and wastewater (17th ed.): Washington, D.C., American Public Health Association, American Water Works Association, and Water Environment Federation, variously paged.

Bonn, B., 2006, 2006 Annual Report of the Tualatin River Flow Management Technical Committee, Clean Water Services and Oregon Water Resources Department, District 18 Watermaster, variously paged, accessed April 24, 2013, at http://www.co.washington.or.us/Watermaster/SurfaceWater/tualatin-river-flow-technical-committee-annual-report.cfm.)

Bonn, B., 2007, 2007 Annual Report of the Tualatin River Flow Management Technical Committee, Clean Water Services and Oregon Water Resources Department, District 18 Watermaster, variously paged, accessed April 24, 2013, at http://www.co.washington.or.us/Watermaster/SurfaceWater/tualatin-river-flow-technical-committee-annual-report.cfm.)

Bonn, B., 2008, 2008 Annual Report of the Tualatin River Flow Management Technical Committee, Clean Water Services and Oregon Water Resources Department, District 18 Watermaster, variously paged, accessed April 24, 2013, at http://www.co.washington.or.us/Watermaster/SurfaceWater/tualatin-river-flow-technical-committee-annual-report.cfm.)

Bonn, B.A., and Rounds, S.A., 2010, Use of stable isotopes of carbon and nitrogen to identify sources of organic matter to bed sediments of the Tualatin River, Oregon: U.S. Geological Survey Scientific Investigations Report 2010–5154, 58 p. (Also available at http://pubs.usgs.gov/sir/2010/5154/.)

Caramujo, M., Boschker, H.T.S., and Admiraal, W., 2008, Fatty acid profiles of algae mark the development and composition of harpacticoid copepods: Journal of Freshwater Biology, v. 53, p. 77–90.

Carter, L.M., Petersen, R.R., and Roe, D.K., 1976, An assessment of the effects of low-flow augmentation and improved sewage treatment on the lower reaches of the Tualatin River during the dry weather season of 1976: Portland, Oregon, Portland State University, Department of Environmental Sciences and Research, 85 p.

CH2M Hill, 1992, Flow augmentation impacts on Tualatin River water quality: Report prepared for Unified Sewerage Agency, Hillsboro, Oregon, variously paged.

Chen, C.Y., and Folt, C.L., 1993, Measures of food quality as demographic-predictors in fresh-water copepods: Journal of Plankton Research, v. 15, no. 11, p. 1247–1261.

Clarke, K.R., and Gorley, R.N., 2006, PRIMER v6, User Manual: Primer-E, Plymouth, UK, 190 p.

Clean Water Services, 2006, Clean Water Services Watershed Monitoring Plan: Hillsboro, Oregon, Clean Water Services, 53 p.

Descy, J.P., 1993, Ecology of the phytoplankton of the River Moselle—Effects of disturbances on community structure and diversity: Hydrobiologia, v. 249, p. 111–116.

Doyle, M.C., and Caldwell, J.M., 1996, Water-quality, streamflow, and meteorological data for the Tualatin River basin, Oregon, 1991–93: U.S. Geological Survey Open-File Report 96–173, 49 p., CD-ROM, (Also available at http://pubs.er.usgs.gov/usgspubs/ofr/ofr96173.)

Fulton, R., 1988, Grazing on filamentous algae by herbivorous zooplankton: Freshwater Biology, v. 20, p. 263–271.

Gulati, R.D., and DeMott, W.R., 1997, The role of food quality for zooplankton—Remarks on the state-of-the art, perspectives and priorities: Journal of Freshwater Biology, v. 38, p. 753–768.

Jeonga, K.S., Joo, GJ., Kimb, H.W., Haa, K., and Recknagelc, F., 2001, Prediction and elucidation of phytoplankton dynamics in the Nakdong River (Korea) by means of a recurrent artificial neural network: Ecological Modelling, v. 146, nos. 1–3, p. 115–129.

Kipp, R.M., McCarthy, M., and Fusaro, A., 2013, *Stephanodiscus binderanus*: U.S. Geological Survey Nonindigenous Aquatic Species Database, Gainesville, Fla., rev. August 28, 2012, accessed May 1, 2013, at http://nas.er.usgs.gov/queries/GreatLakes/SpeciesInfo.asp?NoCache=9%2F19%2F2010+9%3A17%3A16+PM&SpeciesID=1687&State=&HUCNumber=DOnterio.

Lee, K.K., 1995, Stream velocity and dispersion characteristics determined by dye-tracer studies on selected stream reaches in the Willamette River Basin, Oregon: U.S. Geological Survey Water-Resources Investigations Report 95–4078, 39 p. (Also available at http://pubs.er.usgs.gov/publication/wri954078.)

Li, Bo, and Brett, M.T., 2010, Spokane regional wastewater phosphorus bio-availability study—Final report: Department of Civil and Environmental Engineering: Seattle, University of Washington, 68 p.

Mayer, T.D., and Jarrell, W.M., 1995, Assessing colloidal forms of phosphorus and iron in the Tualatin River basin: Journal of Environmental Quality, v. 24, p. 1117–1124.

Oregon Department of Environmental Quality, 2001, Tualatin River subbasin total maximum daily load: Portland, Oreg., Department of Environmental Quality, 165 p., plus appendices, accessed January 24, 2013, at http://www.deq.state.or.us/wq/TMDLs/willamette.htm.

Oregon Department of Environmental Quality, and Unified Sewerage Agency, 1982, Tualatin River water quality 1970–1979—Appendices: Portland, Oregon, Oregon Department of Environmental Quality, [variously paged].

Porter, S.D., 2008, Algal attributes—An autecological classification of algal taxa collected by the National Water-Quality Assessment Program: U.S. Geological Survey Data Series 329, 18 p. (Also available at http://pubs.usgs.gov/ds/ds329/.)

Reynolds, C.S., 1990, Potamoplankton—Paradigms, paradoxes and prognoses, in Round, F.E. (ed.), Algae and Aquatic Environment: Bristol, United Kingdom, Biopress, p. 285–311.

Rinella, F.A., McKenzie, S.W., and Wille, S.A., 1981, Dissolved-oxygen and algal conditions in selected locations of the Willamette River Basin, Oregon: U.S. Geological Survey Open-File Report 81–529, 76 p. (Also available at http://pubs.er.usgs.gov/publication/ofr81529.)

Rounds, S.A., 2002, Development of a neural network model for dissolved oxygen in the Tualatin River, Oregon, in Proceedings of the Second Federal Interagency Hydrologic Modeling Conference, July 29–August 1, 2002, Las Vegas, Nevada: Subcommittee on Hydrology of the Interagency Advisory Committee on Water Information, accessed April 24, 2013, at http://or.water.usgs.gov/tualatin/ann_proceedings.pdf.

Rounds, S.A. and Doyle, M.C., 1997, Sediment oxygen demand in the Tualatin River basin, Oregon, 1992–1996: U.S. Geological Survey Water-Resources Investigations Report 97–4103, 19 p. (Also available at http://pubs.er.usgs.gov/publication/wri974103.)

Rounds, S.A., and Wood, T.M., 2001, Modeling water quality in the Tualatin River, Oregon, 1991–1997: U.S. Geological Survey Water-Resources Investigations Report 01–4041, 53 p. (Also available at http://pubs.er.usgs.gov/publication/wri014041.)

Rounds, S.A., Wood, T.M., and Lynch, D.D., 1999, Modeling discharge, temperature, and water quality in the Tualatin River, Oregon: U.S. Geological Survey Water-Supply Paper 2465-B, 121 p. (Also available at http://pubs.er.usgs.gov/pubs/wsp/wsp2465B.)

Simon, N.S., Lynch, D. and Gallaher, T.N., 2009, Phosphorus fractionation in sediment cores collected in 2005 before and after onset of an *Aphanizomenon flos-aquae* bloom in Upper Klamath Lake, Oregon, USA: Water, Air and Soil Pollution, v. 204, p. 139–153.

Stoermer, E.F., and Julius, M.L., 2003, Centric diatoms, in Wehr, J.D., and Sheath, R.G., eds., Freshwater algae of North America—Ecology and classification: San Francisco, Calif., Academic Press.

Stringfellow, W., Herr, J., Litton, G., Brunell, M., Borglin, S., Hanlon, J., Chen, C., Graham, J., Burks, R., Dahlgren, R., Kendall, C., Brown, R., and Quinn, N., 2009, Investigation of river eutrophication as part of a low dissolved oxygen total maximum daily load implementation: Water and Science Technology, v. 59, no.1, p. 9–14.

Sullivan, A.B., and Rounds, S.A., 2005, Modeling hydrodynamics, temperature and water quality in Henry Hagg Lake, Oregon, 2000–2003: U.S. Geological Survey Scientific Investigations Report 2004–5261, 38 p. (Also available at http://pubs.usgs.gov/sir/2004/5261/.)

Thorp, J.H., and Covich, A.P., 2001, Ecology and classification of North American freshwater invertebrates, 2nd ed.: San Diego, Calif., Academic Press, 1,056 p.

Unified Sewerage Agency, 1992, Flow augmentation in the Tualatin River for water quality enhancement: Planning Division Status Report, 27 p.

U.S. Geological Survey, 2013, Standard Reference Sample project: U.S. Geological Survey Web site, accessed January 29, 2013, at http://bqs.usgs.gov/srs/.

VanDonk, E., and Kilham, S.S., 1990, Temperature effects on silicon-limited and phosphorus-limited growth and competition among three diatoms: Journal of Phycology, v. 26, p. 40–50.

Wehr, J.D., and Sheath, R.G., eds., 2003, Freshwater algae of North America—Ecology and classification: San Diego, Calif., Academic Press, 918 p.

Wetzel, R.G., 1983, Limnology, 2nd ed.: Philadelphia, Saunders College Publishing, 858 p.